O'REILLY® Related

T0252636

Surviving Orbit the DIY Way

By Sandy Antunes
Coming Soon!

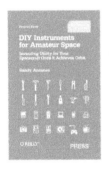

DIY Instruments for Amateur Space

By Sandy Antunes
Coming Soon!

DIY Data Communications for Amateur Spacecraft

By Sandy Antunes
Coming Soon!

DIY
Satellite
Platforms

Sandy Antunes

Beijing · Cambridge · Farnham · Köln · Sebastopol · Tokyo

DIY Satellite Platforms

by Sandy Antunes

Published by O'Reilly Media, Inc., 1005 Gravenstein Highway North, Sebastopol, CA 95472.

O'Reilly books may be purchased for educational, business, or sales promotional use. Online editions are also available for most titles (*http://my.safaribooksonline.com*). For more information, contact our corporate/institutional sales department: (800) 998-9938 or *corporate@oreilly.com*.

Editor: Brian Jepson
Production Editor: Teresa Elsey
Cover Designer: Mark Paglietti
Interior Designers: Ron Bilodeau and Edie Freedman
Illustrator: Robert Romano

February 2012:   First Edition.

Revision History for the First Edition:
        January 27, 2012   First release
See *http://oreilly.com/catalog/errata.csp?isbn=9781449310608* for release details.

ISBN: 978-1-449-31060-8
[LSI]
1327679198

Contents

Preface v

1/Overview 1
    What Is a Picosatellite? 1
    Where Is Orbit? 3
    Pico-Goals 4
        Science! 4
        Engineering! 5
        Art Concept! 6
        The Art of Building 7
    LEO Conditions and Viability 7
    LEO Temperatures 8
        LEO Light 9
        LEO Magnetic Field 9
    How I Got Started 10
    Budgets 11

2/PCB Fab, Soldering, and Electronics 13
    Gerber Files 13
    Breadboarding and Prototyping 15
    Fabricating PCBs 15
        My Criteria for a Good PCB Vendor 16
    Quality Control 18
    Soldering 19
    Parts Is Parts 20
    Parts Suppliers 21
        Rad Hard 24

3/Chassis, Structure, and Gross Anatomy 25
    Parts Is Parts 26
        What the Wizard Gave Us 26
        The Electric Company 26
        The Outer Limits 27
    CubeSats and TubeSats 28
        Materials 29
    Shielding 31

Payload Bays **34**

Flight Spares **35**

4/Power **37**

Power In **37**

Steady State Operations **38**

Burst Mode **39**

Power Architecture **39**

Power Charging Efficiency **40**

Power Out **40**

Power Charging Curves **41**

5/Programming and Coding **45**

Core Processor **46**

Sensor Programming **47**

Packet Data **48**

Radio Transmission Programming **49**

More on Hardware **49**

6/Launching and Rockets **51**

Blowups Happen **53**

Form Factor **54**

InterOrbital Systems **56**

7/Milestones and Planning **59**

Checklists **60**

Research and Design (2 months) **61**

Setup (1 month) **62**

Core Work (5 months) **62**

Integration and Testing (2 months or more) **64**

Team Work **65**

Final Step **66**

Sample Case: Project Calliope TubeSat-style Picosatellite **66**

Assembly Work **66**

Integration Work **67**

Certification Work **67**

Photo Work **67**

Day Job **68**

Goal **69**

Preface

Can any hobbyist build a satellite? Our DIY guide steps you through designing and building a base picosatellite platform tough enough to withstand launch and survive in orbit. If you have basic maker skills, you can build a space-ready solar-powered computer-controlled assembly suitable for attaching instruments and rocketing into space.

Our fundamental premise is that anyone can build a satellite. In Chapter 1, we cover things you can do in space, science and engineering concepts, art/science hybrids, AMSATs, and the potential for advanced concepts such as constellations of satellites. Invent the future!

Chapter 2 discusses the basics of electronics, parts, PCB fabrication, and dealing with suppliers, and has some notes on learning reflow soldering. Chapter 3 then looks at the primary picosatellite chassis that you will use. Choose CubeSats or TubeSats, and you'll find a variety of rigid frame designs, all with the purpose of giving you an instrument bay for your mad experiment.

Chapter 4 discusses satellite power budgets and the limits on solar and battery power, while Chapter 5 provides a quick overview of flyable Arduino and BasicX-24 onboard processors.

To get it up there, you'll need a rocket (Chapter 6), and you'll need to plan then execute your entire build—hopefully aided by the milestone checklists in Chapter 7.

By the end of this book, you should have a strong grounding in the requirements for building a picosatellite that will launch into space. We also recommend the other books in this series: our design, testing and integration book *Surviving Orbit the DIY Way*, designing a mission goal using the power of science with *DIY Instruments for Amateur Space*, and getting your data back to ground with *DIY Data Communications for Amateur Spacecraft*.

In the meantime, I have picosatellites to build! (See Figure P-1.)

Figure P-1. *A TubeSat-style picosatellite being built*

Conventions Used in This Book

The following typographical conventions are used in this book:

Italic
> Indicates new terms, URLs, email addresses, filenames, and file extensions.

`Constant width`
> Used for program listings, as well as within paragraphs to refer to program elements such as variable or function names, databases, data types, environment variables, statements, and keywords.

`Constant width bold`
> Shows commands or other text that should be typed literally by the user.

`Constant width italic`
> Shows text that should be replaced with user-supplied values or by values determined by context.

 TIP: This icon signifies a tip, suggestion, or general note.

CAUTION: This icon indicates a warning or caution.

Using Code Examples

This book is here to help you get your job done. In general, you may use the code in this book in your programs and documentation. You do not need to contact us for permission unless you're reproducing a significant portion of the code. For example, writing a program that uses several chunks of code from this book does not require permission. Selling or distributing a CD-ROM of examples from O'Reilly books does require permission. Answering a question by citing this book and quoting example code does not require permission. Incorporating a significant amount of example code from this book into your product's documentation does require permission.

We appreciate, but do not require, attribution. An attribution usually includes the title, author, publisher, and ISBN. For example: "*DIY Satellite Platforms* by Sandy Antunes (O'Reilly). Copyright 2012 Sandy Antunes, 978-1-4493-1060-8."

If you feel your use of code examples falls outside fair use or the permission given above, feel free to contact us at *permissions@oreilly.com*.

Safari® Books Online

Safari Books Online is an on-demand digital library that lets you easily search over 7,500 technology and creative reference books and videos to find the answers you need quickly.

With a subscription, you can read any page and watch any video from our library online. Read books on your cell phone and mobile devices. Access new titles before they are available for print, and get exclusive access to manuscripts in development and post feedback for the authors. Copy and paste code samples, organize your favorites, download chapters, bookmark key sections, create notes, print out pages, and benefit from tons of other time-saving features.

O'Reilly Media has uploaded this book to the Safari Books Online service. To have full digital access to this book and others on similar topics from O'Reilly and other publishers, sign up for free at *http://my.safaribooksonline.com*.

How to Contact Us

Please address comments and questions concerning this book to the publisher:

O'Reilly Media, Inc.
1005 Gravenstein Highway North
Sebastopol, CA 95472
800-998-9938 (in the United States or Canada)
707-829-0515 (international or local)
707-829-0104 (fax)

We have a web page for this book, where we list errata, examples, and any additional information. You can access this page at:

http://shop.oreilly.com/product/0636920021605.do

To comment or ask technical questions about this book, send email to:

bookquestions@oreilly.com

For more information about our books, courses, conferences, and news, see our website at *http://www.oreilly.com*.

Find us on Facebook: *http://facebook.com/oreilly*

Follow us on Twitter: *http://twitter.com/oreillymedia*

Watch us on YouTube: *http://www.youtube.com/oreillymedia*

1/Overview

All I ask is a successful launch, a clean radio signal, and a life just long enough to achieve that goal.

So you're debating launching your own picosatellite? If high-altitude balloons just aren't high altitude enough, if you feel frustrated by the pace of space development, or if you just really, really like rockets and hardware, I think launching your own satellite is an excellent decision. This book will help you turn that decision into a plan, and turn that plan into finished hardware. But first, what do you want your satellite to do?

What Is a Picosatellite?

Picosatellites, by definition, are extremely small, lightweight satellites. The progenitor of the pico class is the CubeSat, an open source architecture that lets you pack anything you want into the 10cm × 10cm × 10cm cube.

The CubeSat is a satellite as cute as a pumpkin. Forbes reported on one vendor (*http://www.forbes.com/forbes/2010/1122/technology-pumpkin-inc-andrew-kalman-toasters-in-space.html*), Pumpkin Inc., that supplies premade CubeSats. CubeSat itself is a specification, not a piece of off-the-shelf hardware, so Pumpkin decided to prebuild kits and sell them. If you have your own rocket to launch your CubeSat on, for $7,500 they'll sell you a CubeSat kit.

This neatly parallels InterOrbital Systems' TubeSat. InterOrbital Systems (IOS) has the edge in price/performance, as they throw the launch in for the same cost. But it looks like neither IOS nor Pumpkin provide premades, just kits. So there's still hobbyist work involved, but kits remove the need for engineering and just leave the fun part of assembly and integration.

TubeSats and CubeSats are slightly different, of course, and I am insanely pleased that both are advancing the idea of platform *kits*. This is a great step in the commodification of space research. Even if the mini CubeSat looks eerily similar to a Hellraiser Lemarchand box. (See Figure 1-1.)

If you build a CubeSat, securing a rocket to launch it on is not difficult, merely expensive. A typical CubeSat launch cost is estimated at $40,000. There are several commercial providers promising future CubeSat rockets, assuming they complete development. Various NASA and International Space

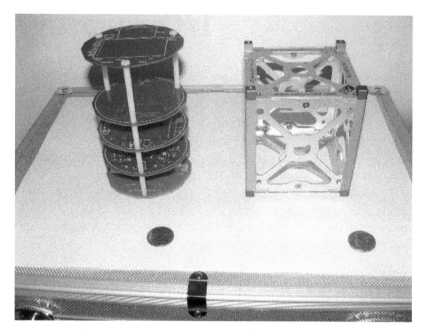

Figure 1-1. *Two variants of a picosatellite, with quarters shown for scale*

Station projects accept some proposals using the CubeSat architecture. There are more companies entering the private launch business each year, so prospects for getting a launch are becoming more robust.

The TubeSat architecture from InterOrbital Systems is an alternative schema. Currently only supported by InterOrbital, it is very cost-effective. You get the schematics, main hardware components, and a launch on their still-in-development rocket for the single price of $8,000. A TubeSat uses a slightly longer hexagonal architecture, 12cm in length and 4cm in diameter.

You can also work with a custom architecture if you have access to a rocket launch (through a college or university, perhaps), but currently the primary two players are the open CubeSat spec and the private TubeSat alternative. For purposes of this book, we'll reference both the TubeSat kit and the Pumpkin CubeSat frame.

Any picosatellite will tend to have these core components:

- An antenna
- A radio transmitter for uplinking commands or downloading your data
- A computer-on-a-chip such as an Arduino or a Basic-X24
- A power system, most often solar cells plus a battery plus a power bus

- Sensors

Where Is Orbit?

First, where will your picosatellite go? It's nearly a given that your picosatellite will go to *low earth orbit* (LEO), a broad band ranging from about 150km up to perhaps 600km. This is the region that also has many science satellites and the International Space Station (ISS). It is in and below the ionosphere, the very, very thin part of the atmosphere that also coincides with much of the Earth's magnetic field.

The Earth's magnetic field shields us from the Sun's most fierce activity. High-energy particles, flare emissions, and coronal mass ejections (CMEs; basically blobs of Sun-stuff) get shunted by the magnetic field before they can reach ground. Where the magnetic field lines dip near the poles, this energy expresses itself as the aurora (see Figure 1-2).

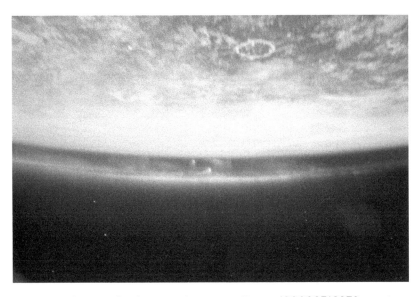

Figure 1-2. *Low earth orbit view of an aurora (image ISS006E18372, courtesy of NASA)*

Above the ionosphere, the space environment can be hostile because of solar activity. Below it, the radiation risks are much lower. This is why the ISS is kept in LEO. LEO is, at heart, about as safe as space can get. It's also where your picosatellite is likely to live.

A typical LEO orbit has about a 90 minute period. That is, it rotates around the Earth once every 90 minutes, doing about 15 orbits per day. Orbits can be positioned near the Earth's equator (equatorial orbits) or loop from the North to South Pole (polar orbits). Similarly, orbits can be nearly circular, or be highly eccentric—coming closer to the Earth at one end of the orbit, and then moving far away at the other.

Your orbit is entirely determined by what your rocket provider has sold you. At the hobbyist level, you're going to most likely get a standard 250km or so nearly circular orbit, either equatorial or polar. Such an orbit lasts (because of drag by the tenuous ionosphere) from 3 to 16 weeks before the satellite will suffer a fiery reentry.

At picosatellite masses, this means your satellite will go up and not return. You have less than three months to gather data. The picosatellite will then, essentially, vaporize neatly upon reentry (no space junk risk!)

Pico-Goals

Given those parameters, what mission goals to tackle? Just what the heck do you want your picosatellite to do? You can neatly break out the typical picosatellite choices into science missions, engineering missions, and art-works. A science payload measures stuff. An engineering payload tests hardware or software. An art project instantiates a high concept. We will visit each.

Science!

On a science mission, your picosatellite will measure something. Science is about measurement at its heart. There are three types of missions you can do: pointing, in-situ, and engineering builds.

A *pointing* mission is like a telescope. Your picosatellite points at an object of interest—the Sun, the Moon, stars, the sky background, or the Earth—and observes it. Note that pointing at the Earth requires a license—not hard to get, but privacy is protected in hobby space.

You can point randomly, but that doesn't seem very useful. You can set a survey mode, where your picosatellite is given a specific orientation in its orbit so that, each orbit, it sweeps across the sky in a predictable fashion. Or, you can do active pointing, making the picosatellite look where you want.

Active pointing is fairly challenging. You need to know your position very accurately. Using inertial references—knowledge of the initial orbit plus internal prediction of how the satellite is traveling—is inexact for sensor pointing purposes. Therefore, pointing typically requires some sort of star-trackers. These are two or more wide-field telescopes that image the sky and compare it to an onboard catalog of known bright reference stars.

Star tracking is technically complex, and likely beyond the weight and design limitations of a typical picosatellite. However, see the section "Engineering!" on page 5 for more on this.

A more common picosatellite science usage is *in-situ measurements*. This is the use of sensors that measure the region the satellite is in without requiring pointing. A thermometer is a perfect example of an in-situ detector. It measures the temperature, and you don't need to precisely point it to know it works.

Other in-situ measurements from LEO can include the electric and magnetic field in the ionosphere, light from the Sun or reflected Earth glow, measuring the ionospheric density, or tracking the kinematics of your orbit and positioning (how you are moving).

Or maybe you don't want to measure something scientifically, you just want to *build stuff*. That's engineering.

Engineering!

An engineering picosatellite uses the platform to try out some new space hardware concepts, or to give you practice in building your own variants of known space hardware.

You can make a picosatellite to test out any of the hardware components. A new power system, a new positioning method, a new type of radio or relay communications, new sensors—really any component of the satellite can be built and improved. (See Figure 1-3.)

Figure 1-3. *Three ounces of flyable instrumentation*

Some picosatellite projects have involved testing—on a small scale—new satellite propulsion concepts, ranging from ion engines to solar sails. Want to test an inflatable space station in miniature, or see if you can make a picosatellite that unfolds to form a large ham radio bounce point? Build it!

Another engineering motive can be to test specific components: for example, comparing a custom electronics rig against a commercial off-the-shelf (COTS) component to see if satellites (of any size) can be made more cost-effective. Or you can test new data compression methods or alternative methods of doing on-board operations.

Innovation in operations is a subset of engineering goals worth exploring further. Picosatellites could be used to test the coordination of a constellation of satellites. They can be test beds for orbital mechanics studies, or lessons in coordinated satellite operations. As the cheapest way to get access to space, they are excellent test beds for prototyping new ways of doing satellite work before moving to million-dollar missions.

Art Concept!

Finally, there are *concept pieces*. My own "Project Calliope" TubeSat (Figure 1-4), used as one reference for this work, gathers in-situ measurements of the ionosphere and transmits them to Earth as music, a process called *sonification*. The intent is to return a sense of the rhythm and activity level of space, rather than numeric data, so we can get a sense of just how the Sun-Earth system behaves.

You can launch a satellite to do anything. Send ashes to space. Ship up a Himalayan prayer flag. Launch your titanium wedding ring into orbit. Any art,

Figure 1-4. *You aren't a real mission until you have your own flight patch*

music, or art/music/science hybrid idea is welcome because it's your satellite. Just give it a purpose or utility beyond just the spectacle of being able to launch your own satellite (Figure 1-5).

The Art of Building

The utility of building is that building your own picosatellite is not just a means to an end, but a worthwhile goal itself. Even if you never launch it, the skills and experience you gain in making your own real satellite can be an awesome experience.

LEO Conditions and Viability

The ionosphere is called that because it is a very thin plasma of electrically charged atoms (ions) and electrons, due to the ultraviolet (UV) radiation

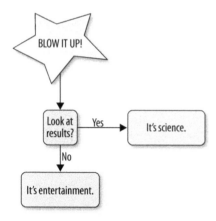

Figure 1-5. *Defining science (courtesy science20.com/skyday)*

from the Sun. Technically it extends from about 50km up to over 1000km (thanks Wikipedia!), but LEO starts at 150km—below that, you can't maintain a stable orbit. The ionosphere, as mentioned, is driven by solar activity. The portion facing the Sun has more ionization; also, solar activity can drive its behavior strongly. There are also dips in the magnetic field line, leading to radiation increases at lower altitudes. We've mentioned the poles, and regions such as the South Atlantic Anomaly (SAA) also have field lines that dip lower.

Having stated you can fly anything, obviously you want to choose what you fly to achieve your goals. For example, if you're sending up sensors, you'll want to ensure a couple things:

1. They have a sensitivity level appropriate to the level of signal you're trying to measure.
2. They have a dynamic range that lets you extract meaningful data.

LEO Temperatures

A metal plate in LEO will cycle from –170°C to 123°C depending on its Sun face and its time in sunlight. If your picosatellite is spinning, this will even out the heat distribution a bit, but that's the range to assume. An orbit has approximately half its time in sunlight and the other half in Earth shade, so the temperature behavior is worth modeling.

Since the picosatellite is spinning, this range is fortunately smaller (as heat has time to distribute and dissipate), and with a 90-minute orbit, you should cycle through three ranges: too cold to register; transition regions where the sensor returns valid, slowly changing data; and possibly oversaturating at the high end. You can add a heater if necessary—satellites have used heaters and coolers depending on the instrument and facing.

Therefore, a thermal sensor (like a microDig Hot brand sensor) that covers −40°C up to 100°C will suffice. The range of −40°C to 100°C is a feasible area to measure. In any event, past that range, the rest of the satellite electronics may have trouble.

LEO Light

Similarly, a light-detecting sensor, for a spinning picosatellite, is likely to return just a binary signal: super-bright Sun in view and Sun not in view. So all that it will measure is the timing of when the Sun is in view. The function of the light sensors will be largely binary, to catch Sun-dark cycles as it spins, as well as the overall day/night cycle of the orbit. If there is a slight tumble to the satellite, all the better. These light sensors will provide a basic measure of the satellite's position and tumbling. If you want to measure actual light levels, your design will have to ensure the Sun doesn't saturate your detector.

LEO Magnetic Field

The ionosphere has a field strength on the order of 0.3–0.6 gauss, with fluctuations of 5%. For a polar orbit, you'll have higher variability and higher magnetic fields than an equatorial orbit (as the Earth's magnetic field lines dip near the poles, hence the auroras). If you want to measure fluctuation, not the field strength, you need to capture 0.06–0.1 gauss signals.

A default Hall effect sensor tends to be designed for Earth work and measures tens of gauss, so you need to ensure the sensor is calibrated for the space environment, not Earth's surface. You can't measure below 0.63 gauss on the Earth's surface, because its intrinsic field will create too much background noise. However, that's not a limit of the sensor. A $10 Hall effect sensor plus an op-amp could measure variations down to as low as 0.06 gauss if there's no large external magnetic field. Below that, the noise from your sensor's circuits, not the sensor, will likely be the limiting factor.

Hopefully this overview gets you thinking about some things you can do with your picosatellite.

On Particle Damage

The mission life is short (less than three months), so you don't need to worry about cumulative damage. I used to do radiation damage models back in school, and it turns out that modern electronics are surprisingly robust on short time scales. You primarily will have single-event upsets (SEPs) that scramble a sensor or computer, but since you likely don't need 100% uptime, this shouldn't be a problem. In fact, glitches will add interesting character to your derived data. Should you encounter, say, a solar storm, it'll be interesting to see how the sensors deal with it, either with saturation or with spurious signals. A proportional counter or ersatz equivalent (like a microDig Reach) can measure these particle counts.

How I Got Started

The Project Calliope satellite is a music/science sonification project that will convert the ionosphere to MIDI data for transmission back to earth. Detectors include magnetic, electric field, thermal, and light. Data is gathered constantly and downlinked as bandwidth is available. Data is sent as unencrypted time-stamped MIDI file packets via amateur radio, transmitted as much as IARU allocation allows. Data usage rights are granted to anyone who can receive it. The satellite will operate continuously within its power and radio budget until re-entry.

A TubeSat picosatellite lifts 200 grams of payload. That's about 7 ounces. Looked at one way, that's less than half a can of soda. But it's enough to lift an entire Nintendo DS game handheld into orbit. Two hundred grams can be a lot of electronics.

When I committed to this project, I didn't yet have the specific electronics in mind. I've built mini guitar amps and guitar sound processors that come in well under 7 ounces. I assumed I could kit-bash stuff and create my own schematics for the final design without much fuss.

In the meantime, I've learned an important lesson. When I started this project, I thought I'd need to assemble a crack team of technical, engineering, and ops people to assist me, while I'd handle promotion solo. It turns out I'll be able to build this satellite with COTS (commercial off-the-shelf) components in a straightforward way relatively alone—but I will need to assemble a team to help me effectively promote it and ensure the music and science results get to people who want them.

What I didn't expect was that there would be companies that already build everything I need. MAKE offers kits that include sensors. I-CubeX in Canada makes sensors for performance art that are IC2-standard science-quality. The "Heliophone" kit alone—a Sun-driven theremin-like device—would make for an interesting payload.

For Project Calliope, the orbital lifetime is brief—6 to 16 weeks, at best. It's intentionally in a low earth polar orbit that will result in re-entry. This means, on the plus side, it's not going to end up as space junk, and it will completely and cleanly disintegrate, becoming lost in the general fall of space dust we already get from nearby asteroids and such. Yes, the Earth gets anywhere from 100 to 1,000 tons of dust each day, just from plowing through space. Which means that every day a few micrograms of dust from space fall on you. Ouch? Hardly.

Budgets

There are five budgets concerning your picosatellite:

Cost
> This is a soft limit because you can always try fund-raising if you exceed your allowance. How much can you afford to spend? If $12,000 is too much, find a cheaper hobby.

Weight
> This is a hard limit. All picosatellites have a capped weight limit that you know in advance (e.g., 1 kilogram entire payload, leaving about 200 grams for the experiment part).

Power
> This is also a hard limit. Your power budget, discussed in Chapter 4, is the amount of power you have via a battery that recharges from solar cells, versus the number of electronics and transmitters you wish to power.

Bandwidth
> A negotiated hard limit, bandwidth is traditionally space-based astronomy's greatest limiting factor. Getting your data down to Earth is yet another budgeted item. You will typically have a limited upload and download capacity because of licensing and spectrum sharing regulations, and you must ensure your data gathering needs fit into those limits.

Processing
> A soft budget is the speed of your onboard computer, which has to be fast enough to deliver your data at the given fidelity to your transmitter, including any real-time onboard processing desired.

Other Orbits

How high is space, how far can you fall with a parachute, where do picosatellites fly, and where does the hard radiation from the sun get nasty? Gathered for the first time in one place is our High Altitude Explorer's Guide.

A typical airplane cruises at 9 kilometers (6 miles) up, around 30,000 feet. Military jets (from the SR-71 onward to modern planes) can hit over 30 kilometers (19 miles) up, over 100,000 feet.

Can you parachute from that height? Yes, in 1960 Joseph Kittinger set the record at 31.3 kilometers (19.5 miles), or 102,800 feet. Felix Baumgartner has assembled a team to work toward tackling a free fall from 36 kilometers up (over 22 miles), an 118,000-foot fall.

But those aren't "space." In the United States, space begins at 80.4 kilometers (50 miles), or 264,000 feet. General international consensus sets a similar limit for the start of space as 100 kilometers (62 miles), or 380,000 feet.

"Low earth orbit" (LEO), where many satellites live, goes from 160 kilometers (100 miles, 525,000 feet) to 2,000 kilometers (1,240 miles, 6.5 million feet). In LEO, we have some sample objects to look at.

Our own Project Calliope satellite will be 230 kilometers up (143 miles, 755,000 feet). The International Space Station (ISS) cruises higher up, from 278 kilometers (173 miles, 912,000 feet) to 460 kilometers (286 miles, 1.5 million feet).

Starting above the *space* limit but a bit before LEO, the inner Van Allen Belts, which magnetically shield the Earth's surface from high-energy particles, extend from 100 kilometers (62 miles, 33,000 feet) up to 10,000 kilometers (6,200 miles, 3.3 million feet).

Geostationary orbits are at 35,786 kilometers (22,236 miles, 117.5 million feet). These geosynchronous orbits, lined up above the Earth's equator, have an orbital period equal to one day, so they "hover" over the same spot of the Earth.

2/PCB Fab, Soldering, and Electronics

In Chapter 1, I discussed the two "cases" or architectures for your picosatellite: the CubeSat or the TubeSat. You will populate these with the guts of your mission. A typical loadout will include the antenna, the radio transmitter board, the solar cells, a battery, a power bus, a processor, and the sensors.

Solar cells will have to be soldered onto Printed Circuit Board (PCB) slats. These will be wired, along with the battery, onto some sort of power bus—another PCB. The on-board processor is likely to be an Arduino or BasicX-24 PCB board. The radio transmitter will be off-the-shelf components wired onto a PCB. Finally, your sensors may or may not have their own PCBs.

That's a lot of boards. Fortunately, fabrication and construction is very easily outsourced these days. You start with a fundamental choice: whether to do through-hole soldering (where you stick the wires for each component through the board, and then solder them with a soldering gun) or to use the lighter Surface Mount Technology (SMT) components.

Most of the MAKE, Arduino, and RadioShack kits out there still use through-hole soldering. It is, in my opinion, an easier task to accomplish. For through-hole, you heat up an iron, place the component, and solder it.

However, the industry is shifting towards SMT parts. They are smaller and cheaper. Soldering SMT requires tweezers, a magnifying glass, and (ideally) the use of a hot plate and heat gun rather than an iron. It requires more dexterity and precision.

Gerber Files

The suggested default TubeSat configuration includes the CAD files necessary to print the Power, Antenna, Microprocessor, and Transmitter PCBs as well as the solar cells. All you have to do is fabricate the boards. There's a standard spec—called *Gerber files*—all fabricators accept.

PCB board designs are typically provided as Gerber files (editable in freeware like KiCAD). These include details for each layer of the board, including

- Where to place the soldering masks (the traces connecting where parts are placed)
- The silkscreen labels that help you know what to place and where
- The location of any drill holes for through-hole components.

 I will emphasize—you can get pre-existing Gerber files for a typical picosatellite project. This frees you from the need to do any electronics design yourself for any parts other than the actual sensors or engineering test design. Put another way, your satellite *bus*, into which you plug your experiment, already exists. This will make starting your picosatellite build very straightforward. There are copious links to CubeSat-ready boards at *http://www.cubesat kit.com/content/links.html*. Remember that CubeSat is a specification, while TubeSat is a specification and a kit; therefore, a CubeSat requires a little more integration effort on your part when designing and building your picosatellite.

There are eight standard files that cover the copper, solder mask, and legend layers plus an overall outline and where you want holes drilled. Make sure you have your PCB Gerber files (aka Gerber RS-274X files), created from a CAD program such as KiCAD. These will include, for a two-layer board, up to eight files as listed below. For a two-layer board, you could squeak by with just three files—you can skip the silkscreen layers if you don't need labels, and you can get by with just one solder mask if it's a one-sided board (no solder traces on the bottom). But all the TubeSat boards are fully defined two-layer boards, so let's move on.

A generic set of Gerber files to order PCB boards includes the following:

1. Top (components side) layer
2. Bottom (copper side) layer
3. Soldermask for top (component) side
4. Soldermask for bottom (copper) side
5. Silkscreen labels for top (component) side
6. Silkscreen labels for bottom (copper) side
7. Board outline (or default is "rectangle that fits everything")
8. Drill holes (.*drl* file, without which no holes will be drilled)

A default single TubeSat requires four different octagonal boards, each about 3.5 inches × 3.5 inches in size, two-layer. It also requires eight 5 × 1 solar panel boards.

Breadboarding and Prototyping

But before we get to logistics, let's prototype. You either are starting with a set of CubeSat or TubeSat reference board schematics, or you are designing your own setup. In both cases, you will want to test each board in isolation before committing to your final build.

I recommend that you use a breadboard or prototype kit for your main processor (again, often Arduino or BasicX-24). These kits include interface software and the programming API. You want to prototype your processor layout, program it in your lab, and then later move it to your flight-ready PCB.

Fabricating PCBs

There are now PCB fabricator companies. You mail or post to them your Gerber files and pay them, and they ship you the boards 16 days later. This is, in a word, awesome. Yes, some makers may prefer to make their own masks and pour out chemicals to etch their own boards, but we are focusing on building a satellite, not smithing your own parts. We are in a golden age of DIY electronics!

My plan was to cheaply acquire good-quality PCB boards sufficient for two TubeSats—the mission satellite and a flight spare. Many companies can turn your plans into small quantities of finished PCB boards, at reasonable cost, in just a few weeks. What is *reasonable cost*? For boards a few inches in size, as low as $25 for a single board. This is the result of my "experiment" trying multiple PCB fabricators. Ultimately I settled on three best candidates and then checked who delivered the best final product.

For my Calliope satellite, I started with the eight printed circuit boards (PCBs) fabricated for the solar cells. These cells are also the main satellite structural beams. The specs: these solar cell panels are about 1 inch × 4 inches. The satellite needs 8 of them, which means I need 10 (2 spares in case I screw up). Oh, actually, I want 20—enough for a flight spare. (See Figure 2-1.)

For those of you too impatient to find out why and how, here's my summary. My favorite for price/performance was *http://PCBInternational.com*, whose DIY special of 100 square inches of board for $99 rocked. I would also recommend *http://PCBNet.com* if you need just one or two boards, as the

Figure 2-1. *Fabricated PCB boards, only $40 each and 5 weeks for delivery*

intro special of $25 per board is very hobbyist-friendly. And a nod to *http://Sunstone.com* for the web interface and reasonable prices; while slightly more costly than the others, the company offered the easiest ordering process for a novice like me.

And now to brass tacks—the whys and hows and who elses.

My Criteria for a Good PCB Vendor

My criteria were the following:

1. Price
2. Ease of ordering/confidence with web-based tools
3. No need to use the telephone. Sorry, if I need to talk to a salesperson to order a board, you're off my list. I'm a hobbyist, not a factory.

If you mix my criteria #2 and #3, you'll also see *fairness*—I wanted a straight price known in advance. Call this *clarity*, perhaps.

I wanted at least two sets each of 4 boards (so I'd have a flight spare), plus 20 solar panels, so my orders ranged from 2 boards to 20, never more. Most came in at $25–50 per board shipped. Really, the price breaks kick in at 10+ or 20+ boards, but I wasn't able to get enough response from other TubeSat builders to really build a group order, so these are my solo prices (see "Batch Ordering" on page 17).

Batch Ordering

The boards get cheaper the more you order. Doing a rough pricing at *http://apcircuits.com*, the cost starts at about $100 for a single set of 10, and drops to $84 per 10 if I order 20. Doing 40 just drops the price to $72 per 10, not a big improvement over the former. But add another 20, and the price drops to $66 per 10, which is a reasonable savings.

If you are teaming with other picosatellite builders, you may want to batch your orders. All I want is 20 inexpensive boards and min-imal hassle. The equation for this is:

```
Savings = (Cost per set of 10)
- (extra shipping involved in a group order)
- (hassle of organizing this)
+ (benefit of not everyone having to create their own fabrication)
```

For the three, here are the prices.

PCBInternational
 3 each of 3 designs for $184 (implying $20 per board)

Futurlec
 2 4 × 4-inch for $47 per board

Sunstone
 $50 per board

This includes shipping, on a 10+ day order (no rush).

The total TubeSat set via PCBInternational is just $370 (that, admittedly, also gives you a few extra spares). Futurlec is competitive—all four main boards, two per, will run you $354 including shipping. Add 20 solar panel blanks for $102 and it's still under $500 ($456) for two full sets of everything you need for a TubeSat. Sunstone was the priciest (8 boards = $400 plus $14 per for the 20 solar panels = almost $700), but I found the web-based interface easy to use, and the company will do quantities as small as 1–2 boards.

A clarification on PCBInternational: they sell a set of square inches, on which you can do three designs per. I'd estimated 8 main boards on a 100-inch-square slab and they gave me 11, so they do a better job fitting than I do! So, all else being equal, one path is to order two 100-square-inch sets from PCBInternational:

- (10 or so) solar cells, plus two copies of two of the main boards
- (10 or so) solar cells, plus two copies of the other two main boards

And for only $35 per slab, you can increase it to 175 square inches each, which nearly doubles your yield and gives you enough for 3–4 satellites. The company has the best price point for this sort of "multiple designs, small quantities" needed, the web interface was good enough, and the shipping time was on par with the others.

Some web discussion boards have reported erratic quality with small-batch runs, so spreading your order among multiple vendors is a good way to ensure against quality control problems. Plus I like shopping.

A dealbreaker was places that said "email for quote," or whose quotes deviated from their advertised rates. For example, one vendor's front page advertised a 4-board minimum order, at $29 per board ($116 total), but their actual quote for a 3.54 × 3.54-inch board says $20.91 per board with a 10-board minimum ($209 total). I'm not sure what the shipping would have been. Since the ordering process is "email us the Gerber files and we'll send back a quote," the company doesn't pass my "novice test," and raises some red flags: quote doesn't match hype, order requires intervention of a sales agent before you know what the final deal is, ordering process not clear, quantity minimums seem inconsistent.

Again, I invoke clarity. If I can't navigate your site, supply my Gerber files, and get the boards at the price you tout, I'm sorry, you're off the list. Extra fees, unclear specifications, or weird quantity breaks will not help you close the deal. A PCB board expert might handle that, but as a DIY-er, I want clarity.

And yes, I accept that "PCB board" is redundant, like saying "ATM machine."

Quality Control

The boards arrive and preliminary inspection shows all boards seem viable, and the alignment seems clean—but I haven't done complete circuit/lead testing yet. I have begun soldering pieces in, though, and am happy to report that the vendors completely failed to screw up (for the ironically impaired, that means "they done good").

Note that if you want to order 10 or more of each board, you can just skip this entire section. Most PCB fabs are very cost-effective at that point. My analysis is only for people trying to minimize their costs for very small (2–3) quantities.

You want to simultaneously order spares (in case you screw up during assembly) yet not buy too much (in case you get a bad batch). Several of these prices might be intro rates or specials and may change in the future. Ultimately I can't prove these are the *one true path* of PCB. But for small quantity hobby ordering for picosatellites, I think these are viable choices.

Soldering

There are many soldering guides on MAKE and on the Web, so we will only summarize it here. The basics of soldering are as follows:

1. Get your iron or heat gun hot.
2. Use heat sinks to isolate heat-sensitive parts.
3. Let the solder flow onto the work.
4. Check the joint; cloudy is bad, shiny is good.

If you are doing through-hole soldering, you will heat the iron, and then hold the iron to the lead you are soldering and apply the solder to the iron, letting it flow onto your joint. Finally, you wick away excess solder.

If you are doing SMT work, you will want to place the board on a hot plate to bring it up to temperature. You will place solder flux onto the pads, and then tweezer the part on top. You use a heat gun with a narrow nozzle to melt the solder and have it flow into place. Finally, you wick away excess solder.

The tools and gear you need include a medium-quality iron. Avoid the dollar-store specials or $10 Chinese eBay specials. You want a decent iron that has variable temperature settings. For the heat gun, choose one with a small aperture, and purchase a set of nozzles that let you narrow the flow of air further. SMT solder paste, consisting of the solder and flux, is usually refrigerated and is perishable.

Be careful with expensive, breakable parts. Primarily, this will be the solar cell wafers you fry. Soldering the solar panels required I learn the new soldering technique of reflow soldering, where you paint solder flux, and then just heat the entire material. I broke two solar cells simply because of the fragility of the cell, not the new technique. Thankfully, IOS provides a few extra cells for this reason.

To understand how solar cells work, the best primer—and it's CubeSat-centric—is up on the useful CubeSat forums at *http://cubesat.wikidot.com/the-technology-of-solar-cells*.

Parts Is Parts

What sort of electronics are we getting into? For the "big three" boards—
Power, Radio, Computer—you are likely to use standard parts. Here's a
rough breakdown on the boards I'm using, with schematics provided by
Gerald Auvray for Interorbital Systems' TubeSat customers.

Power

A couple of high-quality power lithium-ion 3.7V 5200mA cells (fancy
speak for rechargeable AA batteries) and a power bus that includes a
10uH shielded Epcos inductor, 8 MAX9929 current sense amplifiers, a
boost convertor (LM2731SXMF or similar), an ADC (analog-digital convertor, e.g., MAX11112EAP+), a dozen resistors, a dozen capacitors of
varying values, 8 diodes (to go in line with each current sense amp), and
8 Molex headers so you can hook up the other boards to your power
bus. Their kit provides Spectrolab improved triple-junction Triangular
Advanced Solar Cells (TASC), which are noted in one CubeSat forum
(*http://cubesat.ifastnet.com/forum/viewtopic.php?f=4&t=148*) to be
available cheaply.

Radio

A Microhard N920 or TR2M transmitter/receiver plus AFS2 amplifier,
a voltage regulator, a dozen resistors, a half dozen capacitors of varying
values, and several Molex headers so you can get power in and signal
in/out.

Computer

A BasicX-12 or Arduino setup, a 3.57945MHz or similar crystal for CPU
timing, and several Molex connectors so you can get power in and signal
in/out.

Miscellaneous

Solar cells, antenna (typically the metal strip from a tape measure, cut
to size), standoffs, and bolts and nuts.

Ultimately, we're talking less than $300 in general electronics parts (using
2011 prices), plus the fancy pants parts: the BasicX/Arduino ($50–100),
amateur band Radiometrix transmitter/amp pair ($210 + $36), TASC solar
cells ($125), and Li-ion batteries ($75).

To see some schematics for design, a great clearinghouse and starting point
is *http://cubesatcookbook.com/* (though some of the links are stale, there
are links to several of the major players). There are good discussions of specific parts once you start having technical questions at *http://cubesat.ifast
net.com/forum/*.

Parts Suppliers

You can't go too wrong buying from DigiKey in the United States, or Farnell overseas. Be aware that, if you are using someone else's PCB layout, he or she will often provide the parts specifications using the item codes of his or her favorite vendor. There isn't an universal UPC or name for parts.

For most parts, the description is the specification. For example, an Epcos B82464G4103M inductor is an Epcos B82464G4103M inductor, ignoring that DigiKey also sells it under the SKU of 495-3461-1-ND and Farnells' SKU is 742-9444.

Most components such as resistors and capacitors are ubiquitous and nearly fungible. You can swap out manufacturers and minor specs as long as the main desired value (resistance, capacitance, etc.) is fulfilled—and as long as the part has the same form factor. That means "it needs to fit onto your PCB in the holes provided."

The basic parts you will get include resistors (e.g., 47K 5% 0.125W 1206, the latter indicating it's an SMT part) and capacitors (e.g., 4.7uF 16V 1206), which are defined by their electronics properties, independent of manufacturers. Diodes have a short manufacturer-given ID (e.g., CRS06, which is a Toshiba part), and if you have trouble finding one, you can often find an equivalent from another manufacturer.

A fair number of parts will be from a specific manufacturer, including sockets (e.g., Tyco 5-1814400-1) and the occasional IC (e.g., Maxim MAX9929 current sense amp, shown in Figure 2-2). ICs in particular can be hard to find a substitute for if the suggested manufacturer's part is not available, as different manufacturers may use different pin assignments or, worse, different form factors (meaning an alternative part just won't fit on the PCB board as made).

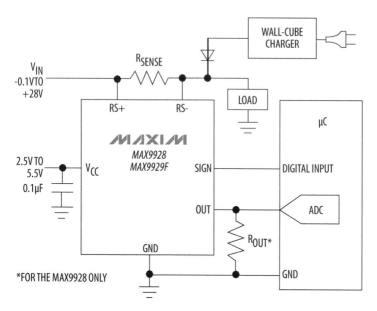

Figure 2-2. *Schematic of a hard-to-find op-amp; always allow for alternative parts!*

Back when I fabricated my PCBs, I thought parts were parts. What could go wrong? My $264 order from DigiKey was short one part, intentionally. The dreaded MAX9929 Current Sense Amp. I needed 8 of these puppies for the power supply part of the satellite. Since I'm building a flight spare, too, I needed 18. Toss in 2 more as spares, and it's an even 20. DigiKey had, in stock, 0. Expected arrival date was January.

Fortunately, another site had them in stock, so I ordered them. Then, later, I got an email from them canceling the order, because they lied, lied, lied about having them in stock (not that I'm bitter...). Okay, not a problem. The engineer who designed the board, Gerard, suggested I try Farnell (his vendor).

Farnell doesn't have a US shop but does run an "export" shop—which was entirely out of the part. Wouldn't have them for 90 days. Hmm. At the time I was not aware of Element 14's Newark shop, which is the US branch of Farnell but has a different catalog. A later check showed they did not have a part by that name. This leads to another issue for the lone hobbyist: which vendor?

Most US hobbyists are familiar with DigiKey, and if not, reading this, you now know of it. There are other electronics shops. Each has different catalog numbers. To some degree, each has different usability issues that can make finding parts hard. Is the amp a "MAX9929" or a "MAX 9929"? A Google shopping search will find absolutely nothing for the former, and find the wrong items for the latter. What is a hobbyist to do?

One avenue is to see if the website that gave you instructions or schematics recommends vendors. Visiting forums for your subcommunity can help—although (as with this) you will run into regional lockouts where a given shop just doesn't sell to your area. There is no one solution.

To bring this hunt to a close, I went direct to the manufacturer. They're out of stock—12-week backorder. Worrisome.

Long story short, after trying a dozen shops, going all the way to Taiwan, I could not find them. I asked Gerard if there was an easy substitute, but he noted that, with my PCBs already fabricated, a similar amp might not fit (or have different pin-outs, which is worse).

Where, oh where, could I find a supply of MAX9929 "Maxim integrated products MAX9929FAUA+ current sense amp, 0.1~28V, 8UMAX, current input bias 1.6µA, input offset voltage 6000µV, current supply 20µA, bandwidth 150kHz, voltage supply 2.5-5.5V"?

A part with similar specs could be substituted, but the question (for the non-electrical engineer) is always which specifications are flexible, and which are not. As a general rule, you can always go with a more generic part—one that has a larger current or voltage range than the specified part. However, you must get a part that has the same pin layout—the same number of pins, the same assignment of which pins receive or put out which signal. Otherwise, you'll have a piece that won't fit your board (different number of pins) or fries your circuit (different pin assignments). Past that, you need to begin a conversation with the board designer to determine which are the crucial specs. Swapping is difficult for the amateur. Offsetting this, though, the hobbyist community includes clueful people who may be able to advise you on alternatives. Failing that, you have to somehow find the original specified part.

Back comes Gerard to the rescue! The French outlet of Farnell has the item, he'll order for me, then ship. All I have to do is pay for parts, and currency difference (minimal via PayPal, fortunately), and international shipping.

Farnell France had only 82 in stock. And Gerard noted that sometimes the company's on-hand is lower than what it lists. Eek!

By my reckoning, I was buying one-quarter of all the Max9929 available to potential satellite builders.

I should have bought them all. Then I'd own the TubeSat Power Management Board electronics market until February 2011, mwah ha ha!

The lesson from this is, once you commit to a given PCB layout, buy all of your more unique parts, especially ICs (the chips), before you place the fabrication order. That way, if you hit a supply problem, you can consider a board redesign before committing. Otherwise, you'll be forced to find a back-alley black-market satellite parts dealer, and face it, there aren't many of us in the book.

Rad Hard

Satellites of old used to use "rad hard" (radiation-resistant) or "mil-spec" (to military specification) parts, as these were seen as more robust and better able to survive orbit. For a picosatellite, though, you can just use off-the-shelf parts. Particularly given the typical short lifetime (less than three months) of an LEO picosatellite, you won't have to worry about damage due to radiation. Your far bigger risks are assembly errors or failure during testing; in fact, we cover that in a separate section.

3/Chassis, Structure, and Gross Anatomy

Did you ever wonder just what a satellite is made of? The answer is... this! (See Figure 3-1).

That's it. Figure 3-1 shows a picosatellite: nearly every piece, excluding instruments and sensors. I have to choose and build those separately. This is the housing, the solar cells and power system, the on-board computer, the little offsets and screws and fairing pieces—everything needed to get me into space.

Figure 3-1. *Parts arrive in a box, looking not much like a satellite*

Let's first look at the gross anatomical considerations for a satellite. The best satellite design will be highly encapsulated, which is to say each subsystem will be independent of the others. A problem with the radio should not affect the sensors, a problem with the sensors should not scramble the CPU, and a program with the CPU should be fixable with a simple reboot.

In mainstream satellite ops, designing autonomous instruments is already the standard. You have the main platform, aka the *bus*, where you monitor health and safety and treat any problem as an emergency (safe hold). If the bus goes down, the entire satellite goes dark. But you want the (often multiple) instruments to be as independent of the main satellite bus as possible. This has two advantages:

- First, it isolates problems. If the bus fails, the satellite is dead either way. But if each instrument is autonomous, the failure of one instrument has no effect on the others, and the mission continues.

- Second, satellite buses are fairly standard, whereas instrument requirements are unique. So it's easier to have a standard bus that you plug in to your one-of-a-kind instrument, than to have to redesign the bus for every new mission.

Fortunately, picosatellites have settled on two standard buses: the existing CubeSat specification, and the fledgling and proprietary TubeSat spec.

Parts Is Parts

Let's zoom in a bit on the TubeSat parts I have, to examine a specific implementation of a satellite's requirements.

What the Wizard Gave Us

Figure 3-2 shows the brains, heart, and courage of the system. Specifically, a microcomputer, a battery, and a not-ready-for-prime-time flight tube. Yes, technically the tube is what the body must fit into, not the actual body—and not *courage*, but that would ruin my Wizard of Oz analogy.

There's also a transmitter there, and I'll be covering data downlink in a separate section.

The Electric Company

Power! More power to him! Err, right, solar power. The satellite uses solar panels to keep the battery charged and perky. Right now, I don't have solar panels, just 48 solar cells (plus 2 spares).

The solar cells came in a Styrofoam rack (see Figure 3-3), as they are highly fragile. As if the Styrofoam packing wasn't warning enough, I also received

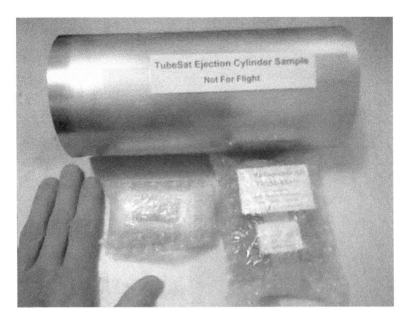

Figure 3-2. *Close-up of primary components for a TubeSat platform (sans instrumentation)*

warning emails from IO Systems reminding me to be careful. Part of the reason these are fragile is they are not in their final form.

To be useful, they must be attached to something stronger, and of course wired up. I received the PCB board schematics by email, and fired up a program called KiCAD to read them. I have eight of the boards fabricated and shipped to me. I'll slap the cells onto the board, wire them up, and we'll have bona fide solar panels.

The Outer Limits

Ever build something and end up with a lot of extra parts left over? Whoops. Okay, these aren't really extra parts. These are the pieces that don't go onto the actual satellite, but help with building and programming it. Cables and a USB interface board to hook it up to a PC, CDs with documentation, and a little cable that I never figured out, probably a sensor cable for an non-I2C sensor.

Figure 3-3. *Solar cells, very fragile*

CubeSats and TubeSats

We start with a picture (Figure 3-4) of a CubeSat skeleton (ordered from Pumpkin) and a TubeSat skeleton (created using PCB specifications from InterOrbital Systems).

In terms of material, a CubeSat uses an open or closed metal case. You manufacture and attach your satellite PCBs and sensors with metal bolts. A 1U CubeSat will have a Base Plate Assembly, a Chassis Wall Assembly, and a Cover Plate Assembly.

A TubeSat reduces weight by using the satellite PCBs as the structural elements, separated by Teflon spacers.

Both CubeSats and TubeSats can be bought as preselected kits. Direct comparison is slightly nuanced. A Pumpkin CubeSat containing the skeleton and basic processor and I/O runs $7,500, with $5,500 or more additional for adapters and modules. A TubeSat kit (parts only, no launch) is $750 and includes the microprocessor and transceivers, but it does not include the PCBs; those you fabricate for perhaps $500 more.

For those of you in Europe, the CubeSatShop (*http://www.cubesatshop .com/index.php?page=shop.product_details&flypage=flypage.tpl&product _id=1&category_id=1&option=com_virtuemart&Itemid=66*) sells CubeSat

Figure 3-4. *Again, our two variants of a picosatellite, with a quarter for scale*

skeletons for €2300. Clyde Space (*http://www.clyde-space.com/cubesat _shop?gclid=CMbX4ejboKsCFSM65QodBmlhfA*) sells the ISIS 1U CubeSat for £2250. Back in the United States, Astrodev (*http://www.astrodev.com/ public_html2/*) has CubeSat shells for $1250 and also lists transmitters, antennas, and other components.

Additional vendors appear about every two months, so at the time you are reading this, I hope you have multiple potential companies willing to service your satellite anatomy needs. CubeSat Cookbook (*http://CubeSatCook book.com*) has some partial lists of the growing number of vendors.

It is worth noting that you can adapt the hexagonal TubeSat PCB Gerber files to a square CubeSat configuration to create an excellent prototype for your CubeSat; anyone doing this exercise is heartily asked to please contact this author so we can disseminate your contribution to the community!

Materials

PCB quality is (as covered in Chapter 2) high. PCBs are made of ceramics and plastics, and will experience outgassing when in vacuum. Outgassing is when volatile chemicals and, basically, junk evaporate off a component once it is in vacuum. Certain materials outgas more than others; liquids have strong outgassing, Teflon has very little.

Your satellite will have a primary outgassing when you test it in a vacuum chamber on Earth (to be covered in a later volume in this series). If possible, you want to let each component outgas in isolation, as all that outgassing material will settle on something. In particular, you want to make sure any sensors do not have outgassing material condense and form a residue upon them.

So bear in mind, any sloppy excess flux, any dust, and any usage of off materials (such as PVC instead of Teflon) will come out in vacuum. Fortunately, you can clean and prep your assemblage before vacuum testing.

Satellite assembly is usually done in a clean room (see "Clean Rooms" on page 30) for this very reason. With picosatellite work, however, although DIY clean rooms exist, I found them difficult to use. I visualize DIY satellite work as being very dynamic and fraught with experimentation as to the best design. Further, the final package is small but not compact—and therefore easy to clean.

However, if you are using or testing optics (rather than in-situ sensors) or have moving parts, you may have to use a clean room. Optics require cleanliness because any smudge or debris on the mirror, lens, or focusing apparatus will appear in every data frame. Moving parts must be clean because any grit or dirt, once in vacuum, will lose any lubrication due to humidity or oil and quickly freeze your moving system; a vacuum is very hostile to moving parts.

Clean Rooms

A *clean room* in rocket parlance is exactly what it sounds like—a very clean room. Typically, this means taking a room that is spic-and-span, and creating positive pressure. You make sure that clean air is being pumped into the room, and rely on the fact that the air has to leave as your way of keeping dust out.

A common setup is to run powerful intake fans that are well filtered, so they pump and filter lots of outside air in. You surround it with plastic curtains, so the air can escape, and rely on the mild wind (for want of a better term) to ensure no outside dust enters.

For bonus points, you can make the scientists wear bunny suits— that's the actual name of the white clean room outfits. There are varying degrees of clean room standards. You can find plans for a "DIY Clean Room" that uses simple materials (a storage chest,

---

gloves, and plastic) at the Instructables (*http://instructables.com*) site. For our work, we aren't doing fine optics or detectors with a high contamination risk, so we need only a moderate system.

Over at MAKE, they have something more my speed—a DIY cleanbox (*http://blog.makezine.com/archive/2006/10/how_to_make_clean_box_low.html*).

It's just what it looks like—a plastic tub covered in transparent plastic, with little rubber gloves inside that you access from holes on the outside. But even that may be overkill for our picosatellite, because, again, we have no optics, sensitive electronics, or other easily contaminated items.

In previous NASA missions, I've seen flight hardware sitting in the middle of dusty labs. Most hardware is fine that way, because most of it is a big hunk of metal. The only time for clean rooms is final assembly, when you put in the sensitive optics. You do vacuum tests later, and outgassing usually removes most contaminants. Vibration testing, likewise. Wrap it up and you're done.

Cleanliness is a progression, not an absolute from square one. So first I'll just be in a clean, cat-free room. Later, I'll use the trunk to transport to the metal shop. Then back to the lab. I'll be building a vacuum chamber and vibration rig for testing. After that, I'll transfer it to a sealed bag for final shipment. Our result will be a clean, but not ultra-clean, functional satellite.

It is up to you what cleanliness fidelity you require; however, I submit that a well-kept, clean work area in which you only take out a PCB or component when you are working is sufficient. Keep accurate and clean control of the parts, store them in a safe, dry case when not working, and use alcohol and compressed air to keep the parts clean, and you should have a clean-enough mission able to survive vacuum testing and, ultimately, orbit.

Shielding

In general, picosatellites are assumed to have minimal shielding. There are two main uses for shielding the electronics and instrumentation: preventing your satellite from being noisy and protecting your satellite from space.

The first is necessary: shielding your electronics so they do not put out radio frequency (RF) noise that contaminates your instrument measurements, your own radio transmitter, or other satellites nearby. Given the low power levels involved, the last bit (contaminating past your own satellite) is unlikely and such noise would be a result of very poor parts choice or assembly, rather than being a shielding issue. Shielding for internal noise may require specific grounding or small bits of foil covers, but neither introduces a serious weight factor.

Shielding your electronics from space, whether it be from the varying ionospheric interactions in orbit or from catastrophically large solar (space weather) events, is more problematic. The first problem is that shielding requires weight. The second is that shielding is subtle. Incident particles (charged electrons and protons) can do transient and long-term damage to your electronics. However, transients can be recovered from, and a typical picosatellite will not be in orbit long enough to accumulate long-term damage.

More critical is that shielding is not a simple matter of adding material to make you safe. Damage occurs when charged particles or EM fields penetrate the surface layers of your satellite and electronics to interact with the inside of your parts, especially the inside of any integrated circuits. However, space has a spectrum of energies. Low energies will be blocked by very little material. High energies will pass right through your satellite without interacting. The primary damaging particles are in a middle range, where they are partially slowed (attenuated, or shifted) by your satellite such that they deposit all their energy into your satellite, rather than being fully stopped or passing entirely through.

Shielding blocks low energies and attenuates the higher energies to lower energies. You therefore need to know the distribution of energies—how much flux there is at low, medium and high ranges. In a worst case, you might add shielding, only to find that the bulk of the spectrum you are interacting with is being attenuated to that *sweet spot* where it causes the maximum interference with your device.

I recommend that you check and remove any RF noise your satellite corrects. If you feel your electronics are particularly sensitive, you can look up the spectrum of expected radiation and particle counts by wavelength and their equivalent attenuation coefficients to do a complete analysis of radiation risks. The most likely case is that, while a little foil shielding will cut off the bulk of the low energy, use of heavy shielding should only be done if you are willing and able to do this sort of radiation analysis, and in most cases, you will find heavy shielding brings no benefit relative to its weight.

Autonomous Satellites

How smart does a satellite have to be to function? It is entirely possible to design a satellite that doesn't require a programmable CPU. Like B.O.B. from *Monsters vs. Aliens*, it doesn't need a brain. All it needs is energy, sensors, the ability to yell or shut up, and a payload.

Let's back up a bit. My picosatellite kit has a computer core and a Radiometrix transmitter. I unpacked it and then discussed the satellite's gross anatomy earlier in this chapter. But how minimal can I go?

Power is pretty basic. Solar cells charge batteries. Sensors and the transmitter use power. Since we're always powered up (unless we run out of power), I don't need fancy power management, just "run until you die." So no brains needed for that.

I need the Radiometrix transceiver to send down data. And HAM requirements include being able to turn off the transceiver if so ordered. So I need a "smart switch" that recognizes the signal for *off* or *on*—some sort of analog-to-digital (A2D) circuit, such as a N0QBH Packet Radio Decoder. Minimal brains needed.

The sensors already send out data as I2C, and I just need to feed them into an appropriate HAM digital signal. So I need a signal processor, but no additional computing needed—no clock signals or timing, no on/off, nothin'. Just run the sensors to the transmitter so (if the transmitter is on) it gets sent out. If the transmitter is off, it doesn't matter what happens, so I don't need to deal with handling that.

Yes, an Arduino or similar processor can send or receive I2C, but that's overkill. I don't need to *do* anything to my MIDI data, just send it down as HAM packets. So I need a modem or TNC (terminal node controller) that basically converts the I2C to AFSK tones.

You can make the computer unnecessary, and just use a few low-power circuits. However, I accept it's more efficient from a "time spent developing" perspective to use a generalized CPU instead. Generalized CPUs are cheap, easy to program, and handle everything a custom circuit would.

That's what makes this science. It isn't what my opinion is, or what I like better. It's all about what a) works and b) works best, where *best* is quantifiable and measurable—the least power, highest stability solution.

Payload Bays

It is important that you design your picosatellite to include room for the sensor or engineering payload. While this is obvious, it is so easy to get caught up in building efficient power systems and programming the CPU, only to realize you don't have enough space for your mission.

Imaging sensors—sensors that require a line of sight outside the satellite shell—require the most difficult placement. In-situ sensors, which measure the satellite's ambient environment, must be placed where noise from the onboard electronics do not contaminate the signal. Therefore, isolation and shielding are considerations, which will affect both your spatial and weight budgets.

Shielding is a complex topic. Here, we are concerned about shielding your components to prevent cross-talk—to prevent any of your components from causing interference with the others. This is something you can and will test for when you integrate your pieces into your final build.

There is also shielding from the external environment expected in space. LEO is in the ionosphere and there are radiation and electric/magnetic field concerns. Unfortunately, shielding against those are not a simple matter. First, shielding requires adding material, which adds weight.

Second, while high-energy particles will do more damage than low-energy particles (typically), electronics are susceptible to specific energies more than others. Finally, shielding will block low-energy particles but it will also slow down (or attenuate) higher-energy particles to a lower energy. It is possible, through incorrect design of shielding, to increase the radiation hazards experienced by your satellite.

For both the weight and complexity issues, then, I recommend that you focus on shielding in terms of ensuring your satellite components will not interfere with one another. In particularly, you do not want noise from your power bus or CPU to interfere with your sensors, and you want your radio antenna to be on the side opposite any electromagnetic sensors to minimize induced radio interference. Past that, skip the weight hit and fly it naked.

Flight Spares

Let's close with the idea of flight spares. The idea here is twofold. First, it is good to have a second satellite ready in case a mishap occurs to the first. Mishaps can range from *rocket blew up* all the way down to a mundane *dropped it while carrying it to the truck*.

Conceptually and more important, you want to build two or three satellites simultaneously for two reasons. First, you may make a construction mistake with one. Having a spare means you can continue work without having to wait for new parts or fabrication.

Second, you will build one better than the other. Statistically, one of your builds will have better performance than the other. This better one is the one you will fly. By creating multiple builds, you give yourself and your skills a chance to practice, hone, and ultimately create a better picosatellite.

So build two and fly the one that does best in tests.

4/Power

Power is one of the primary budgets for your satellite, and is a hard limit that must be obeyed. If you exceed your bandwidth budget, you lose some (but not all data), but if you exceed your power budget, your entire satellite cannot function.

Fortunately, the calculations for your power budget are relatively straight-forward. Solar cells will charge the batteries so they are X full on average, with a minimum power of Y (during dark periods, at maximum predicted discharge). Your satellite will use Z power.

- If $Z < X$, you live.
- If $Z < Y$, you thrive.

Power In

There are two parts to your power calculation. First and foremost is your maximum and average solar cell power capability. This sets the limit on your energy budget. You can have 10kg of batteries that can potentially store 3 Amps of power, but if your solar panels have a maximum total load of 310mA (per second), you will never have more than 310mA sustained power available.

Put another way, you cannot charge batteries with power you don't have. The batteries exist only as a storage device to keep excess solar power during the night portion of your orbit. Your power is solar.

Unless, that is, you aren't using solar (see "Alternative Power for Satellites" on page 37).

Alternative Power for Satellites

Satellites use solar power because sunlight in space is cheap, free, lightweight, and ubiquitous. While a gasoline-powered satellite isn't feasible (no air for combustion), some missions have used thermal nuclear reactors (RTG) as a power source. It is unlikely you can get permission to fly radioactive power plants at the DIY level, however, as even NASA has to jump through many regulatory hoops for theirs.

Being DIY means thinking outside the box, however, and power is just another box that could use a few new openings. It might be possible to fly a hydrogen fuel cell into space. Although it requires consumable fuel, rather than the free eternal sunlight, it has potential for short-term high energy usage or for habitat testing.

A typical fuel cell can convert solar-generated electricity plus onboard water into hydrogen and oxygen. Operated in reverse, it can convert hydrogen into power. If you are flying a mission that can use oxygen or that requires water, experimenting with fuel cells can be a neat engineering project.

As the ionosphere is magnetically charged, in theory you might be able to generate an induced electric field from the passage of a metal satellite through a magnetic field. Bottling this ionospheric energy—which is at a very low level—again could be an interesting engineering experiment.

We fall back to our original justification for solar: cheap, lightweight, ubiquitous. But in DIY, remember the best way is never the only way, and you may discover something surprising.

Let's look at the TubeSat reference architecture:

- Solar cells: (48) 2.5V, 62mA
- Lithium-ion 3.5V, 900mA

An estimate of maximum solar charging is (ignoring voltage conversion for the moment) simply the number of cells times their output. 62mA * 48 cells yields 1488mA. Predicting that your orbit will spend half the time in sunlight and half outside, and that only one-third of the satellite will have a line of sight to the Sun at any point, you have a reasonable expectation of an average charging ability of 250mA.

For a rule-of-thumb figure, assume that one-sixth your total solar panel output will be available as an average power during the entire mission.

Steady State Operations

Your power budget is affected by your anticipated power usage profile. If you intend for your satellite to be on most of the time, with a sustained consistent power usage, you typically want usage Y to be at or under one-half your solar capacity X.

Under that *steady stage* plan, the satellite operates using half the flowing solar power while the sun is shining, while the other half of the solar power charges the battery. During dark, the battery fully powers the satellite and slowly discharges, but it does not reach zero before the next sun passage.

This is a typical plan for a sensor-driven or communications satellite mission. You are on at all times, and periodically dump your data back to Earth.

Burst Mode

An alternative mode is *burst mode*, where you charge the satellite but do little to no power usage for most of each orbit. Your satellite will have a minimum power usage: processor on, radio transmitter able to receive signals, heaters (if needed).

Then, when there is enough accumulated charged battery power, you can engage in a higher-energy experiment, such as testing an electric ion propulsion system or deploying a folding structure.

This model requires either direct commanding from the ground or very intelligent programming for autonomous operations that takes into account power available. It is most suitable for engineering projects that exceed your expected *per second* solar power input.

Power Architecture

You have five parts for your power system architecture. The solar cells provide power and charge the batteries. The batteries provide power. The power bus is a PCB that provides the interconnections between *things giving power* and *things using power*. Finally, you have *things using power*—the satellite processor, transmitter, and sensors.

You can structure this architecture using several circuit paths.

Direct charging
> Solar cells → batteries → power bus → satellite/instruments

In this configuration, the solar cells are always charging the batteries. Everything operating on the satellite is always draining from the batteries. This is the easiest configuration to design and is very robust in operations.

Managed power
> Solar cells → bus → (satellite, instruments, batteries)

If you want a more intelligent power bus, you can run the solar cells into the bus, then have it choose whether to charge the batteries (if they need it) or supply excess power (past battery charging) directly to the instruments.

This provides less wear and tear on the batteries and ensures no overcharging, but requires a more sophisticated electronics design. However, if you are operating in *burst mode*, this alternative layout provides more accurate power level monitoring because you have separated your power sources (solar cells and batteries) at the bus.

As available battery-free operations
   Solar cells → satellite/instruments

This mode is rather risky. Your satellite is only on when the sun is shining. When the orbit goes dark, so does the satellite. You save on weight, as you don't need to carry batteries. If you are doing a concept piece or art/science project, this could be viable. Your primary concern is that you have no communications capability during the dark, and your onboard processor will be rebooting once each orbit—turning on only when solar power is available. While not recommended, it is an option.

Power Charging Efficiency

Fellow TubeSat pioneer Wesley Faler of Fluid & Reason has calculated power curves we can expect for our orbiting picosatellites. He has generously allowed me to publish my summary of his analysis. His summarized estimate is that a six-cell solar panel in a sun-synchronous polar orbit with perfect positioning can expect to produce 0.5 watts, or achieve approximately 15% sustained charging efficiency. This sets our ultimate power budget for the satellite, and helps us choose appropriate instrumentation and control schemes.

Power Out

A BasicX-24 requires 20mA itself, plus up to 40mA per I/O channel used. An Arduino (*http://arduino.cc*) requires 50mA, plus the same 40mA per I/O channel used. The bulk of sensors used will be passive, which is to say they require no additional power past that provided by the I/O connections. Finally, a Radiometrix UHF transceiver uses 27mA to receive and 110mA to transmit a 100mW signal; boosting it to 500mW (e.g., Radiometrix AF2S amplifier) requires 2mA when receiving but a hefty 250mA additional power when transmitting.

We can now calculate our power budget, as shown in Table 4-1.

Table 4-1. *Power budget*

CPU	50mA
Radio (waiting)	30mA
Sensors (assuming four)	4 x 40mA = 160mA
Radio (transmitting)	750mA
Power usage, nominal	240mA
Power usage, transmitting	990mA

Recalling our "power in" calculation, we anticipated a 500mA average power availability. What we first realize is our nominal (gathering data and open to receiving signal but not transmitting) power rate is half our average power, and we are safe.

The second realization is that we don't have enough power to be in transmission mode all the time. For our typical DIY picosatellite, this is acceptable. We have *burst usage*, where for brief periods we will exceed our average power, but on a per-orbit basis, we can still come in under budget.

Max transmission time calculation: 90 min × 500mA power in, minus 90 min × 240mA nominal usage, yields (23400mA ÷ 750mA transmitter usage) yields 31 minutes per orbit available for full-power, happily transmitting operation.

So the situation isn't too bad. Using these initial profiles, we can still transmit for up to one-third of a given orbit and remain within our expected power budget.

You can run the numbers yourself very easily for alternative configurations. This is an essential activity well worth spending a day on, as your power budget is a fundamental entity that you cannot exceed.

Power Charging Curves

For those who are interested in a more thorough analysis of how the solar cells charge over an orbit, we can go into depth using the aforementioned Wes Faler analysis. His simulation and calculation, running in full about five pages, is an excellent breakdown of the issues involved. His own note on it mentions "the calculations have been updated to use the AM0 space spectrum with a power density of 1366.1 W/m^2 rather than AM1.5 spectrum (1000 W/m^2) quoted on the solar cells for ground operation." I'm presenting a summary of his analysis here, both to provide specific information on TubeSats and, more generally, to show how to analyze and simulate orbital conditions for a project.

Let's break down his work into pieces. First, he sets up the physical units for the solar cells and for the orbit:

- photoWidth = width of photocell panel (meters) = 0.02109
- photoRadius = distance from center to face of photocell panel (meters) = 0.04343
- photoAngle = angle of circle for one photocell panel (radians) = 27.295 × 2π/360
- alAngle = angle of circle for one aluminum strip, radians) = 17.632 × 2π/360
- photoLength = length of the photocell panel, meters = 0.127
- wattPerPanel = 2.277 cm^2/cell × 0.27 efficiency × 0.1366 W/cm^2 × 6 cells/panel = 0.4928
- halfOrbitSeconds = seconds for half of an orbit at 310 km = 45 × 60

Of particular interest is that the watts per panel comes out to just under half a watt for an ideal cell (of normal efficiency). The problem to solve is whether, in orbit, we can actually achieve this. Remember we start with his conclusion—a six-cell panel can produce a half watt. So this suggests we operate at about one-sixth efficiency because of orbital issues.

What Wes considers is, first, the panel positioning. After setting up the geometry, he rotates the panels around the TubeSat center and determines, relative to the line to the Sun (placed along the +X axis), which panels get any sunlight. He notes "these must still be multiplied by the power per solar panel and then integrated over time as the satellite orbits."

Rotating the TubeSat along the X-axis (sun line) yields no power change (obviously; the panel angles and coverage relative to the sun do not change). The power factor for rotation around the Y- and Z-axes do vary the power received by the panels, as shown in Figure 4-1.

Wes then solves for three possible orientations. *Flat orientation* is where the TubeSat's long axis is always facing the direction of travel (the *bullet* configuration, I'd call it). As it moves through its orbit, it goes from nose-toward-Sun (at one pole) to broadsides at the equator, down to tail-toward-Sun (opposite pole). For flat, "2042 Joules are taken in during the half orbit facing the sun. Over the half orbit period, this equates to an average of 0.75 Watts."

Radial orientation is where the TubeSat's long axis always points toward the Earth. The satellite is broadside to the Sun at each pole and has its endcap face the Sun at the equator. Much as with Flat, "2047 Joules are taken in during the half orbit facing the sun. Over the half orbit period, this equates to an average of 0.75 Watts."

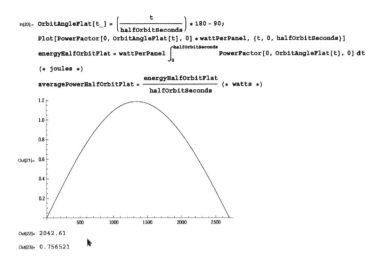

In[20]:= OrbitAngleFlat[t_] = $\left(\dfrac{t}{\text{halfOrbitSeconds}}\right)$ * 180 - 90;

Plot[PowerFactor[0, OrbitAngleFlat[t], 0] * wattPerPanel, {t, 0, halfOrbitSeconds}]

energyHalfOrbitFlat = wattPerPanel $\int_0^{\text{halfOrbitSeconds}}$ PowerFactor[0, OrbitAngleFlat[t], 0] dt

(* joules *)

averagePowerHalfOrbitFlat = $\dfrac{\text{energyHalfOrbitFlat}}{\text{halfOrbitSeconds}}$ (* watts *)

Out[21]= (plot)

Out[22]= 2042.61

Out[23]= 0.756521

Figure 4-1. *Orbital power curve for a rotating satellite, showing charging versus orbit (image courtesy and copyright Fluid & Reason)*

Closing off our analysis by Wes, we have *SunSeeker orientation*, where the TubeSat is always broadside to the sun and "likely rotating along [its] long axis for cooling." For this power-optimizing orbital orientation, "3208 Joules are taken in during the half orbit facing the sun. Over the half orbit period, this equates to an average of 1.18 Watts."

The actual satellite orientation one choses will be based primarily on the sensor requirements. Where you need to point and how you have to face is the primary driver for this. However, once you decide on an orientation, power considerations can kick in. You can rely on the default orientation—simply tumbling and requiring no specific pointing. In our later discussion of sensor loadouts and missions (covered in *DIY Instruments for Amateur Space*), we discuss different pointings. Once you have a mission pointing, you can run the above power curve calculations.

But pragmatically, assume you will have 15% the maximum potential charging or less on a typical orbit and your power budget will be reasonable.

5/Programming and Coding

Potentially, your processor can control everything from maneuver and power through to data gathering and transmission. You do have a processor budget—how many operations per second it can run. You also want a bulletproof mission, which suggests as little processing as possible to ensure that your programs are simple, debugged, and only used when needed.

Therefore, I recommend you not use computer commanding of your power system, but rely on good power architecture. Your primary use of the processor is for three purposes: data handling, turning your radio on and off, and commanding any engineering tests.

Data handling
  Process and optionally store onboard sensor/project data.

Turning the transmitter on or off
  • Receive ground commands telling satellite to broadcast its data to Earth.
  • Send data to the transmitter for broadcasting down to Earth.
  • Process any *turn off now* instructions.

Commanding any engineering tests
  Only if you have an engineering experiment—such as an ion drive—that you wish to turn on, turn off, or change the operation of during the flight. Most sensors, in contrast, will be "always on," even if there isn't bandwidth to transmit.

It is better to have the sensors always on, requiring no ground commanding, and throw away data if you cannot transmit it, than the alternative. The alternative is to use valuable radio time to tell the satellite to do its work. Design your satellite sensors to be always on within your power budget and you remove the need to have to code them to turn on or off.

Core Processor

Your DIY picosatellite will need a brain. But which brain? IOS's kits include the BasicX processor; for Christmas I received the Arduino kit so beloved by DIY folks. Both are potentially flyable. Let's compare.

BasicX-24 (*http://www.basicx.com/*): 32K memory, requires 20mA plus up to 40mA I/O loads, operates at −40°C to +85°C. Programmed in BASIC (ugh) via serial cable. See Figure 5-1.

Figure 5-1. *BasicX-24 processor and computer reference board (image Copyright NetMedia, Inc. 1998-2011)*

Arduino (*http://arduino.cc*): 32K memory, reported typical <26mA current draw plus 40mA per I/O load, operates at −40°C to +125°C (estimate based on range its optional temperature sensor functions at). Programmed in a C subset via USB port.

So we end up with a classic rocket science trade-off. There's an easier to use, robust kit—the Arduino. And a more bare-metal rig—the BasicX—that has better power usage but is a bit harder to hook up. Both are flyable. So do I go for ease of use or performance?

I'll clearly go for performance. The lower power requirements of the BasicX are an overwhelming plus. As a computer scientist, I'm not worried about the ground issues—hooking it up, programming it, testing. This is a first launch, and I need to ensure everything is as tight as can be.

However, were I to launch a second satellite, I'd be tempted to try the Arduino to see if it is an easier architecture to develop for. And in fact, I am working on a CubeSat using Arduino as a future project.

For any architecture, you can get into power tweaking. Our maker editor notes that "you can get pretty hard-core on this," such as with the JeeNode Arduino clone that runs at 3.3V and 9.8mA draw (see *http://jeelabs.org/ 2009/05/14/power-consumption-baseline/* and *http://www.sparkfun .com/tutorials/309*), not counting the use of a watchdog timer to get under 1mA draw by sleeping 99% of the time. For our designs, we are assuming the CPU must always be available, if only to receive potential FCC radio shutdown requests, but judicious use of sleep can result in power savings.

Sensor Programming

Although we discuss sensors in depth in the third book in this series, *DIY Instruments for Amateur Space* (O'Reilly, 2012), you have to think about your sensor programming architecture when you are choosing your build-out for your basic satellite platform. Sensors will be analog or digital. Analog sensors require you hook to the A2D (analog-to-digital) ports on your microprocessor, while digital sensors will hook in directly through digital input pins.

There are competing standards for hobby sensor use. A fixed-baudrate serial line (UART) is not something I recommend, as one thing you require is to identify a given sensor as well as receiving the data. For a single sensor, however, it is possible to go with a serial connection. CAN Bus or 1-wire, similarly, are technically possible and have some library support for both processor types, but I would not recommend them.

I2C is a low-powered two-wire sensor standard that can run at a 10kbit per second low-speed mode using 3.3V or 5V signals, ideal for the processors chosen. Standard libraries exist for both BasicX-24 and Arduino to manipulate I2C data. The primary advantage of using I2C is the wealth of premade sensors that use this format.

Data typically comes in constantly, but to save on storage, you can decide whether to transmit periodic values, rolling averages, or deltas. For a periodic value, you check the incoming data stream against your processor's internal clock and sample the data at the desired interval, discarding the rest of the data. A rolling average adds the current value to the existing average and either transmits, stores, or simply remembers that value until such time as transmission is ordered. Delta values do not measure the actual sensor entity (for example, measuring *temperature is 69°C*) but instead tell you how much the value has changed since the last reading (again, measuring the change in temperature instead, for example, *temp changed by 1°C*).

The use of delta values versus absolute values, and sampling versus averaging, are driven by the science needs of your mission, covered in depth in *DIY Instruments for Amateur Space*. At a programming level, you need to primarily concern yourself with the *acquire data, process, store or transmit* flow you will need.

Packet Data

Your processor will send data to your radio transmitter chip. Radio transmitter chips can use the amateur (HAM) bands, such as the IOS-recommended Radiometrix TR2M, or the Industrial, Scientific, and Medical (ISM) band (with a Microhard n920 or Microhard n2420). Your transmission strength is likely limited to half a watt—but what to send? The answer is packet data.

Packet data is a way to send digital information (such as image data or sensor readings) using analog radio. Your ground station will need to be able to handle packet data, directly interfacing with a computer. Your satellite will simply connect your Arduino or BasicX-24 to the transmitter chip.

Transmitter chips are smart; if you send them the appropriate-sized packet with necessary headers, they transmit. Therefore, as a minimum, you will need to bundle your data into a data packet containing your ID, a timestamp of when the data was taken, any sensor ID required that lets you know which instrument was used, and the actual numeric value of the data. You need to process all incoming data and decide which to transmit. Amateur HAM uses the AX.25 data link layer specification to define its packets; however, within that specification, you can use any underlying data representation you wish.

Technically, you can maximize your effective bandwidth by ignoring data timestamps and broadcasting just data values, with the time then being assumed to be sequential and close to the time the data was received. This would be the equivalent of real-time streaming data. However, data packets perform best as timestamp/data value pairs, because then you can reassemble the packets on the ground regardless of how acquired or how complete the acquisition.

A rough workflow of your data handling code is as follows:

```
raw data from multiple sensors
  -> add time stamp and detector ID
     -> apply any onboard processing or calibration
        -> filter which data to transmit
           -> send to transmitter
```

On the data end, it turns out the biggest barriers aren't technical but regulatory. That's probably no surprise to most on this list. A half-watt Radiometrix transmitter plus amp is standard for amateur satellites and has a good track record. The puzzle you raise about capturing enough music to make it worthwhile isn't a power budget issue, but an IARU/ham radio spectrum sharing issue.

Radio Transmission Programming

The FCC rules the spectrum, but the International Amateur Radio Union (IARU) is the entity that actually coordinates satellites. If you are using amateur band, you need to file/coordinate with the IARU to use the amateur band with your personal ham callsign as the satellite's callsign. Like any regulation, there are many details.

The main IARU requirement is *play nice, and be able to turn off your transmitter at a moment's notice.* My approach will be to have the required *stop transmit* command, of course. To make this really work, however, I will also have all transmissions time out and shut down automatically after an appropriate interval (10 minutes of no contact). Transmission only starts when a *start transmitting* command is uplinked, with possibly a few orbits that are prearranged to have the satellite automatically turn on based on a clock (if IARU allocates the time).

Since most hams use transceivers (radios that can receive and transmit, for non-hams reading this), it is more fair to only broadcast when someone is willing and able to receive and yet shouldn't reduce the ability to get pico-satellite data down.

From a ground station listener point of view, when you wish to listen, you transmit the *start transmitting* command to the picosatellite; then it broadcasts for the set time (while you're in range) and then shuts down. By definition, you can also shut it down earlier (since you're the one that sent the *on* and are still in range), so it provides assurance that your picosatellite won't hog spectrum. This complies with the IARU requirements without requiring me being able to guarantee 24/7 uplink to the satellite. Note that the IARU is an association, not a protocol or rule set, however, so any approach is evaluated by them on a case-by-case basis to determine what is acceptable.

More on Hardware

Your CPU limits are given by the hardware. A typical BasicX-24 is a 7.37MHz CPU with 32K bytes of user program and data storage, has a maximum program length of 8,000 lines, and has 400 bytes of working RAM (providing a maximum executable program size). It has two serial ports (1200-460.8K baud) and 16 standard I/O pins that can also be used as 8 10-bit analog inputs (ADCs) or 8 digital I/Os. It runs at 5V.

An Arduino, depending on the variant, has 14 standard I/O pins, allowing for six to eight 10-bit analog inputs. The Arduino Mega has many more I/O pins, but it also draws more current than other Arduinos. Most Arduino variants use digital pins 0 and 1 for a serial line, but the Arduino Mega has four serial lines. The standard Arduino (Arduino Uno) has 31.5K of effective storage and 2K of RAM. Arduinos run at either 8 MHz (requires 3.3V) or 16 MHz (requires 5V).

A good satellite design will not push the limits of either processor for the simple reason that your payload must be designed to be minimal, stripped down, and very robust.

6/Launching and Rockets

Picosatellites have launched on such a mix of rockets that it is hard to categorize a single approach. Most AMSAT and university projects ride piggyback as an extra tiny payload on an existing space mission. These rocket launches are negotiated by people with access—people who know other rocket-launching people. NASA, Japan's JAXA, and other space agencies occasionally run launch opportunities for picosatellites. CubeSats, despite becoming a standard architecture, do not have a standard launch provider. Like any picosatellite, they rely on who you know to launch. Some have launched via Russian converted ICBMs, others lofted up to the ISS, still others use the piggyback method.

Now that we are entering a new commercial space age, we hope access improves. The above methods historically involve costs from $50,000 to $150,000 for just the launch fee (according to CubeSat co-director Jordi Pulg-Suari).

NanoRacks LLC (*http://nanoracks.com*) likens itself to a "no-frills airline" and is brokering CubeSat launches up to the ISS from $25,000 (for a 1U CubeSat for an educational client), with commercial clients being charged a higher rate of $50,000, and non-US clients being charged even more.

The checklist from the NanoRacks website lists some interesting services worth evaluating for any launch provider:

- Paperwork required for space transportation
- Handling of the safety review
- The space transportation to the space station
- Insertion of the payload into the NanoRacks Platform
- Power
- Date return

Note that since NanoRacks uses the ISS, it can return the satellite. JAXA is planning to deploy CubeSats from the ISS as well. Most rocket launches, in contrast, will be a one-way trip, with your payload experiencing fiery reentry within three months (tops) after launch.

SpaceX announced pricing (*http://unreasonablerocket.blogspot.com/ 2011/08/thoughts-on-small-sat-and-cube-sat.html*) of approximately $70,000–100,000 for a 1U CubeSat put into low earth orbit (LEO). The Russian Dnepr converted missiles have done multiple CubeSat launch attempts (*http://www.spaceandtech.com/spacedata/elvs/dnepr_specs .shtml*), with polar and equatorial orbits ranging from 200km to 800km and prices of approximately $10M for a 2900–4500kg payload (200km) or a 150kg payload (800km). This requires you work with other satellite launchers to fill a roster, but provides the lowest price-per-kilogram (under $1,000 for 200km altitude or $10,000 for 800km altitude, assuming a 1kg payload).

Nanolauncher wants to use converted jets (*http://www.technewsdaily .com/recycled-military-jets-serve-as-satellite-launchers—1494/*) to achieve LEO insertions. ISILaunch offers to negotiate launch space (*http://www .isispace.nl/index.php?option=com_content&task=view&id=51&Itemid= 81*) for you and any of the rocket startups that competed for the X-Prizes are worth tracking to see who succeeds.

The lowest fixed-price offering out there is InterOrbital Systems (see Figure 6-1), offering 1kg TubeSat launches for $8,000 (including a TubeSat kit) or a 1kg 1U CubeSat launch for $12,500. The company is still building toward its first launch, however.

Figure 6-1. *InterOrbital Systems CPM mobile launch rail (image Copyright InterOrbital Systems 1996-2011)*

It's almost silly to call the existing CubeSat community "the establishment," but indeed there is an existing channel for projects (particularly educational ones) that want a broker to help them reach space. Easiest to reach among these is the CubeSat coordinator group (*http://www.cubesat.org/index .php/about-us*) at Cal Poly. As one helpful reader put it, "You mention that you have to know people to get on a flight. If you build a CubeSat you do know someone, the coordinating group. Cal Poly has been organizing flights and as long as you have a CubeSat built to the standard and the money to pay for the launch you can fly. They have set up the relationships with the launch providers as well as coordinate the qualification tests the providers require. They typically have several launches being coordinated at any given time so you can pick the one that is at a good time frame and CP can provide the milestones and requirements to make that launch" (anonymous but verified comment at *http://www.science20.com/satellite_diaries/cubesat_eating _tubesat-83273*).

Cal Poly as a coordinator is excellent at reducing the technical barriers, but there still are costs associated with a Cal Poly launch. In some ways the group is a cost-savings middleman, but I have not yet found free rockets for hobbyists. The CubeSat coordinating group is an excellent resource, but for some DIYers, the soft requirement that you negotiate with people (rather than simply paying for or commissioning a provider) can also be a barrier. The requirements say "all you need is a project that conforms to the CubeSat spec," however; so as long as there are enough rockets, there is hope.

"Focus: CubeSats—A Costing + Pricing Challenge" (*http://www.satmaga zine.com/cgi-bin/display_article.cgi?number=602922274*) by Jos Heyman is an in-depth analysis of both work costs and potential launch costs. His analysis sets formal launch preparation at $16,000, and a low $1,000 actual launch fee, presuming you've negotiated a free "goodwill" ride on an existing booster.

We are at a cusp. Currently, the way to launch your DIY satellite seems to be *talk to people and find someone with some spare rocket space*, which is perhaps the least effective way to run a project. However, several companies seem to be offering potential launches in the sub-$20,000 range, if they can get their technology together.

Blowups Happen

It is entirely possible that your original launch vendor may become unavailable. Even with the CubeSat standard format, this can leave you without a rocket and facing a possible interface/integration quandary if your new launch provider uses different specifications or has different driving requirements. If you go with an alternative sole-supplier form factor such as TubeSat (typically to save on cost and overhead), your risk if you shift launch providers can increase as well.

I am launching my Project Calliope picosatellite on an InterOrbital Systems (IOS) rocket. IOS invented the TubeSat format. What if InterOrbital fails—its rockets all blow up, it runs out of money, it decides to do interpretive dance instead of rockets? Is Calliope dead?

This is especially important because I did fund-raising. So (besides my ego) other people became invested in the project's not-guaranteed success. Bear in mind the fund-raising gave them actual cool stuff—mission patches and flight pins. Stuff is nice. Stuff is tangible. The flight patches will remind them that they supported an independent space venture, whether it blew up or not. Also, given my tenacity, odds are good I'll find a way around setbacks. But I do not want to be known as the *it blew up* mission.

The question remains—what if your rocket provider fails? There are two types of failure conditions likely:

1. They go out of business and can't launch you. Solution: find a new provider.
2. Your rocket blows up upon launch. I recommend a refly.

Form Factor

In this book, I present examples using both the CubeSat and the TubeSat standards. If you have to switch providers and are going from CubeSat to CubeSat or from TubeSat to TubeSat, obviously no form factor change is needed. However, what if you have to shift from the proprietary TubeSat design to a CubeSat specification?

Geometrically, a TubeSat is a cylinder 9cm in diameter and 12.7cm long (see Figure 6-2). A standard 1U CubeSat is 10cm × 10cm × 10cm. Therefore, a TubeSat, if trimmed, can fit neatly into a CubeSat. The main body of the TubeSat is easy to shorten simply by decreasing the size of the payload bay. I feel the bay is generous already—you can fit an entire guitar pickup in there. You can also tweak the spacers for the other section to push them to a closer tolerance.

The TubeSat solar panels are slightly more difficult to trim; there is perhaps 1.7cm of excess PCB you can safely cut without needing to refabricate. That still leaves a TubeSat solar panel at 11cm. To resolve this, it is simplest to create a new solar panel PCB that fits only 2 × 2 rather than 2 × 3 solar cells, and fabricate enough boards to fit all the cells. Since the surface area of the 10 × 10 × 10cm CubeSat is larger than the 10cm tube, there is plenty of extra room to place the additional panels.

In terms of complexity, the main TubeSat PCBs (power, transmitter, computer, your payload) are complex, while the solar panels are dirt simple. Also,

Figure 6-2. *Size and weight build model for a tubesat-type 1kg limit picosatellite*

the bulk of your satellite building time will be spent getting the payload sans panels to work. The solar panels, in terms of construction, are nearly an afterthought, added on after everything checks out. Therefore, of all the parts to modify, the solar panels are the most trivial to resize.

The TubeSat2Cube mod starts with defining your CubeSat shell with the resized solar panels. You then attach your working, tested TubeSat payload within that cube. The extra height lost to the payload area can (much as with the solar panels) be made up for with the extra volume the CubeSat provides.

Conclusion: if IOS fails, just launch the TubeSat as a 1U CubeSat. It'll be more expensive ($40K and higher launch fees are typical), but it requires no payload modification; lets you make use of all the design, development, and build effort already spent; and costs less than $400 in additional parts. As other companies enter the launch fray, a CubeSat launch cost should drop.

All risks can be mitigated if you plan in advance, and you must always be willing to adapt to the hazards and joys of our new age of commercial space.

InterOrbital Systems

With more than 24 "we plan to launch" commercial rocket startups entering the business, and with only 2 (Scaled Composites with *SpaceShipOne* and *SpaceShipTwo*, and SpaceX's *Falcon*) getting serious press time, it is difficult to describe a generic rocket profile. However, we can look at InterOrbital Systems' to view a detailed briefing of one modular approach into low earth orbit.

IOS's modular rocket system (the *Neptune*) is about to test a one-module Common Propulsion Module (CPM) as a sounding rocket/flight test. The company has a completed rocket, mobile launch rail, and FAA permit. All its rocket plans use one or more CPMs, bundled for the mission needs. One CPM is a suborbital (ballistic) launch, but IOS is not interested in suborbital.

The company's stance is that suborbital is not a natural progression to orbit; it can be a dead end. Orbital is where it's at. Noted IOS's CEO, Randa Milliron, "I can't stand to see things stop at suborbital, like it's the be-all end-all definition of space flight. It's fine as a start, if you consider moving up from that."

The one-CPM "SR145" will list 145 kilos to 310km ballistic. Bundle together seven CPMs, you get the *Neptune*, doing low earth orbit (LEO) missions. More, you get the Moon. But we're getting ahead of things.

Right now, the company uses a mobile launch system. Tests are currently north of Mohave, in the Mohave Test Site. Tests use a FAA Class 3 waiver, so it's not a fully fueled launch. Upcoming orbital launches will require an over-ocean permit. The current Class 3 tests will include payloads from orbital clients who want to test communications systems (and I would presume, flight resiliency).

IOS, the FAA, and just about everyone likes over-ocean launches, as there are fewer liability issues. For ocean-type launches, condensation and icing and valves freezing is also minimized. IOS is using multiple facilities, and looking at using Wallops or Kodiak Island for launches, too.

Fuel is a primary consideration with rockets. You need a lot of thrust to lift a heavy weight against gravity. Fuels are typically rated by their specific impulse, or Isp, the rocket equivalent of *miles per gallon*. A fuel can have a high efficiency—such as a pulsed plasma or ion engine, which accelerates hydrogen atoms at high velocity to provide a kick. But some high efficiency engines do not produce a large thrust, they simply produce an efficient push--much as electric cars are deemed highly efficient regarding fuel but low performance as engines. A good rocket fuel requires an efficient engine that has high efficiency but can also generate a large total thrust. This means both a good Isp and also a sufficient mass of fuel to provide decent thrust.

For fuel, IOS rockets use white fuming nitric acid, turpentine, and furfuryl alcohol. Sounds nasty, but it ends up being both cheap and ecological. Since turpentine is derived from pine trees, the rocket consumes eco-friendly and renewable trees, rather than nonrenewable hydrocarbons.

The white fuming nitric acid, specifically, is pure and concentrated, 99% pure, bought in trucks along the highway, a standard industrial product—not exotic, fairly cheap (as well as safer). It's similar in price to liquid oxygen (LOX). It has moderate performance, high density, easy storability, easy availability. In short, it's a great oxidizer.

As a bonus, they can pump it out of the rocket back into storage if a mission scrubs (unlike LOX).

The turpentine is basically the regular paint-store product. Turpentine is also a renewable fuel. Its Isp is 240 sec or so at sea level (using nitric acid), and it's slightly denser than kerosene. Furfuryl alcohol is used for ignition—it's made by Quaker Oats.

For cost, fuming nitric is about the same cost as LOX, forty cents a pound. Turpentine is less than gasoline. Only a few pounds of furfuryl are used, at around $3 per pound. So all three are very economical.

In contrast, nitro tetrox is around $25 a pound, other hybrids can be higher (mostly because only one company makes the hydrozine-based propellents), and they're also environmentally difficult. IOS doesn't use kerosene because it wants to stick with easily stored fuels and avoid cryogenics, plus kerosene doesn't burn smoothy with nitric acid.

For fuel handling, IOS uses a closed pressurized system to transfer fuel to avoid contact with the air. Nitric acid is an irritant, not poison like red fuming nitric acid (which includes nitrogen tetraoxide). Protective clothing and breathing apparatus are used in case of leak/vapors, but those are more benign than, say, hydrozine or nitrogen tetraoxide. Full fuel specs and stats are at IOS's rocket test page.

For the test/orbital flight timeline, Randa notes, "Everybody thinks they'll have their flight six months earlier," with Rod Milliron adding, "Our pace of development is based on our funding stream." The company's current milestones are based on events and achieveables, not dates—dates only get people in trouble. Historically, even with the motivation of million-dollar X-Prizes, no launch company has really ever operated under any schedule more accurate than "we will launch when we're ready."

IOS emphasizes being a small company. Randa says, "I'm the CEO and I help make the ablative liners, I apply composites. Everyone is involved here...it's as lean as you can get, everyone here is hands-on and enthusiastic and that's the way it should be." She adds, "Those haters out there...they know what they can do."

IOS's Rod Milliron invented the TubeSat kit, the famous $8K satellite plat-form + launch that Project Calliope favors, the kit that uses off-the-shelf parts. Rod's TubeSat is a kit he designed to work with COTS parts. IOS has also added CubeSat kits, at $15K including launch. Randa's closing words apply to us all: "Don't ever let anybody stop you."

7/Milestones and Planning

Schedules are better than plans. Just blocking out the time during which you will tackle a task is more important than figuring out how you're going to do it. We're going with an assumption that you are motivated and competent. Everything else is logistics and being a quick study on details.

For a large project, you need to have a plan because you have multiple people involved. Even in an Agile setup, you need a project plan. "Agile" isn't an excuse for no planning or for laziness. However, once that's set up, it's too easy to fall into the trap of making plans, to-do lists, and task orders—and find the work has slipped past you.

Whereas, if you are scheduled to work on a task, you will find yourself actually getting it done. For my Calliope build, the only reason I made continuous forward progress was by blocking out time. This meant that I'd always a) post an update on Tuesdays and b) spend some time at my homemade lab bench.

That I always posted at *http://Science20.com* on Tuesdays meant I had to have something to post about, which gave me an incentive to keep the project moving forward. Forcing myself to be at the lab bench (small as it might be—Figure 7-1) meant the satellite build always progressed.

This wouldn't have happen if I kept reading things, making plans, strategizing, developing concepts, and similar useful—but not generative—thought work. Night is for thinking, day is for doing.

Many others have built small satellites, but (to my knowledge) always as teams, and frequently as teaching exercises. I am pushing the use of COTS (commercial off-the-shelf) parts, but that too reflects a trend that is increasing in labs everywhere.

In any project, momentum is just as useful as certainty. You build on the stuff you know and, ideally, gather up colleagues or a team so that you can solve problems as the situation gets more complex.

Some might see this as ill-planned, but it's actually a good process when doing inventive work into unexplored terrain. I'm a a one-person basement music/art project designed to be operational—not just *it flew* but trying to deliver a useful product to a community.

In short, I have a plan, but the final details are in motion.

This is a good thing. If I had needed every answer, complete and set in stone, before beginning this project, it wouldn't be novel or cutting-edge. Much of the fun of DIY (do it yourself) is learning to do it.

And as it happens, my approach is similar to the US Army troop leadership procedure. Theirs goes:

1. Receive the mission.
2. Issue the warning order.
3. Make a tentative plan.
4. Start necessary movement.
5. Reconnoiter.
6. Complete the plan.
7. Issue the complete order.
8. Supervise.

As noted in the book *High Altitude Leadership*, "The beauty of this system [is that] you don't complete the plan until you've started the necessary movement and reconnoitered the battlefield."

That said, having checklists of expected and indeed necessary steps from *idea* to *rocket* is helpful. The rest of this chapter provides you with the steps needed. You will have to assess how much time each step will take you and your team, given your access to tools and your skill sets.

Keep forward momentum going, and you will succeed.

Checklists

You should give yourself one to two years to create your picosatellite from concept through to launch. At the same time, the actual build work could be done as quickly as two months, followed by two months of integration and testing. Reconciling these disparate numbers is the purpose of this chapter.

Starting backwards, the build will take just one semester, one summer, three short months or so. However, this step is like the stuffing in an Oreo cookie: it's the tasty middle surrounded by the rigid cookie frames of your entire project support system.

Put another way, building is easy if you have a clear design, all the parts, a well-stocked workshop, and a chunk of time freed up to do it. To get to that

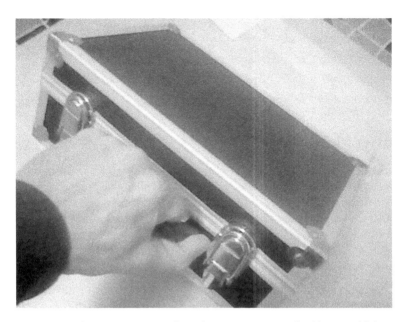

Figure 7-1. *An entire picosatellite plus parts remains highly portable!*

point, you have to do the more unpredictable steps of research, design, team-building, and gear acquisition. After you build, you will have to do programming, integration and testing, and actually shipping your picosatellite to your launch company well in advance of the launch date.

Here is a rough schedule, assuming you are spending 10–20 hours a week on your picosatellite:

Research and Design (2 months)

- Concept and design
    - — Choice of picosatellite template
    - — Desired rocket
    - — Discussion
- Partners
    - — Recruiting of initial team
    - — Choosing launch vendor
        - — Going from research to roster

- Design
  - Choice of chassis
  - Sensors: I2C? General analog-to-digital? Other?
  - Custom boards for power, communications, processors?
  - Acquire Gerber files for circuit boards

Setup (1 month)

Setup, aka "shopping," is a key component because you will have to wait for parts and fabrication orders to be completed and shipped. The lead time needed to get a PCB board is typically 18–30 days. That is, you wait up to a month from when you submit the order to have the boards in hand. Similarly, ordering electronic parts will have a shipping delay, plus perhaps additional delays if any parts are back ordered. This is why setup is broken out as a separate item.

It is best to get everything you need before you begin, so you are not wasting valuable build time waiting on supplies. Once you've placed your orders, you can then use the shipping window as a period to learn some necessary skills, such as picking up a ham license or learning SMT soldering techniques.

- Prebuild
  - Research and choose vendors
  - Make parts lists
  - Make list of test gear needed: borrowable/rentable versus must make yourself
  - Determine telecom licensing needed
- Workspace
  - Establish workspace
  - Set up with furniture (desk or bench, lighting, ventilation)
  - Assemble or buy tools, especially for electronics work
  - Security: Lockable versus easy access?

Core Work (5 months)

- Core satellite build (sans payload)
  - Creating schematics
  - Ordering PCB boards and parts for chassis and main boards.

- Sensors work (the payload)
    — Designing and testing sensors
    — Creating sensor test lab
    — Choosing final sensor mix
- Core satellite integration
- Sensor integration
- Programming
- Checklist
- Telecommunications
    — If amateur band, obtain ham license
    — If regulated band, begin licensing
    — Contact IARU once launch is set
    — Build low-power ground station in test bench
    — Determine and acquire final ground station equipment
- Build and buy for processor (processor reference board)
    — Buy reference kit
    — Set up with laptop/PC
    — Learn editor and file transfer utilities (*hello world*)
- Buy and build for sensors
    — Option 1: Test sensors using COTS interface
    — Option 2: Test sensors using processor reference board
- Primary buy schedules: flight and flight spare
    — PCB fabrication outsourcing versus fabbing yourself
    — Chassis and structural: prebuilt versus custom design
    — Platform PCB boards: power, communications, processor
    — Antenna: design and fab (or part of kit)
    — Instruments: ordering parts
- Assembly
    — Build dummy skeleton (full-scale reference model)
    — Build full, real hardware, no-electronics mechanical model
    — Solder all PCBs with full electronics, test in isolation
    — Complete any mechanical parts

Note that actual soldering of any single board typically takes 2–4 hours, depending on the complexity of the board. You'll often want to do that in two segments—first all SMT parts, then finishing with the through-hole parts. You also want to clean the board of excess solder and cruft. You want to test the traces to ensure it is clean, which can easily take an additional hour. Finally, you want to do this for two boards for each component—the main and the flight spare. For a given satellite with 5 main boards, then, you are looking at soldering, cleaning and testing 10 boards. Assuming you can only spare 10 hours a week, that suggests completing one pair of boards each week is a reasonable goal.

- Programming
    - Telecommunications
    - Sensors calibration
    - Sensors readout
    - If using processor to monitor power system (not recommended) code it
    - Other

Integration and Testing (2 months or more)

- Integration
    - Power trio: solar, battery, bus
    - Power→transmitter
    - Power→sensors
    - Power→processor
    - Sensors calibration
    - Sensors→processor (same as reference architecture)
    - Transmitter→antenna→testbed radio
    - Processor→transmitter
- Testing
    - Power budget (full on, nominal usage, quiet)
    - Ability to reboot
    - Communications
        - Telecommanding, especially transmitter turn on/off
        - Advanced telecommanding, if any (including uploading new code)
        - Downloading of sample data
    - Sensor sensitivity, calibration

- Deployment
- Vacuum and initial outgassing
- Thermal vacuum
- Vibration/shake
- Drop
- Ability to reboot (again)

Team Work

This is a very time-consuming component, often considered "overhead," that surrounds your mission, and is hard to account for as a line item. This includes learning what you need to do, talking to outsiders about your work, keeping your team together and motivated, and handling a variety of licensing and regulatory issues.

- Team building
  - Finding people
  - Assigning roles and benchmarks (or deadlines)
  - Motivational needs: events, tchotchkes
- Marketing and funding
  - Building awareness:
    - Getting Press coverage
    - Making a website
    - Using social media
    - Blogging the build
    - Souvenirs
    - Trade shows and (un)conferences
    - Photograph needs (during entire build)
  - Funding
    - Angels
    - Kickstarter
    - Patrons
    - Selling t-shirts and souvenirs
    - Yourself (cash or credit?)
- Illness, attrition of team

Final Step

- Ship to launcher (and storage of flight spare), ideally two months before launch.

Sample Case: Project Calliope TubeSat-style Picosatellite

I am often asked questions about my musical picosatellite, Project Calliope. Easy questions have concrete answers. "What are your sensors?" I-CubeX magnetic, thermal, light. "What magnetic field is expected?" About 0.3 gauss. "How are you going to distribute the tracks?" As free remixable MIDI files via Web.

Others are either vague or awkward. "When will the satellite be done?" Obviously "by launch." "What will it sound like?" Whatever the musician wants. "What's your downlink bandwidth?" It's shared bandwidth, so I only have estimates.

"I don't know yet" is a scientist's favorite phrase. It means we're in motion and doing cool stuff, with a wide open future ahead. There is no shame in admitting you don't know something. The only shame is if you stop there, if you decline the chance to explore.

Into every satellite a little grunt work must fall. Today you get to read the exceedingly boring but entirely real details of a typical week of satellite construction and project management.

Assembly Work

To build the satellite, I have to solder electronics parts onto a PCB board. This is a straightforward task requiring only a modicum of coordination, and I've been doing the soldering work in fits and starts over the past months. However, I hit a snag with soldering parts. My tried and true pistol-grip soldering iron has become bent, so I picked up a new pencil-type iron. Alas, on a test project, the new iron failed. I ended up with solder "whiskers" on some of the connections. Whiskers is an ugly and known case of bad soldering. So I need to get a new iron and test it out before I risk further soldering on real Calliope parts.

I've been holding off putting the actual radio transceiver and amp on the radio board, and the BasicX-24 onto the computer board, because those are the only components that I do not yet have duplicates of (due to their high cost). I don't want to screw them up. Still, I would like to put them in this week or next. Ideally, I can finish one set of all the boards to begin integration and testing.

Integration Work

Integration is often the hardest and most detail-oriented part of any mission. In theory, I have all these circuits and sensors and pieces that work okay in isolation. Putting them all together and ensuring they talk to each other—that'll be the trick.

In particular, I had to tackle the BasicX-24 programming in earnest. I had tested the sensors earlier using the Midi rig to ensure the sensors worked. That means the sensors were hooked into the ICube-X-recommended interface and then hooked into a laptop computer.

However, for actual flight, I needed to hook them up to the BasicX-24 board that will actually fly, and that is substantially less powerful and less intuitive than using the full-blown laptop with GUI-based software that I tested under. It's sort of like switching from camping in a mobile home to backpacking with a tent and sleeping bag. Sure, it's all rough living, but one is more bare bones than the other.

To jump-start the programming, I have the IC2 library and sample code for the BasicX-24, but I needed to get it all working with the sensors in earnest. I also needed to ask my satellite expert (Gerard) for his sample radio code. Reusing existing code is the name of the game here, not just because it's easier, but because I'd rather use code that works and has been tested, than create stuff from scratch.

Certification Work

As far as when Calliope will launch, InterOrbital has been chugging forward with its test flights; it has FAA permits for its first test suborbitals. Calliope, being orbital, flies after the company finishes verifying its design with these "up then down" flights. InterOrbital has a strong "we will fly when we are ready to fly" ethos, so an exact flight date isn't nailed down. I did need to file for radio permission with the IARU in advance, so there was a bit of guessing on launch dates (on my part) to proceed with the legalities needed.

Photo Work

My spouse briefly became enamored of setting up a new photography rig for documenting the Calliope build process. Her approach would be better than my current "two lamps and a digital cam" approach, no doubt. However, in all things I favor "getting it done" over "doing it better," so I continued documenting the build using the earlier way (Figure 7-2) until the new photo rig was ready.

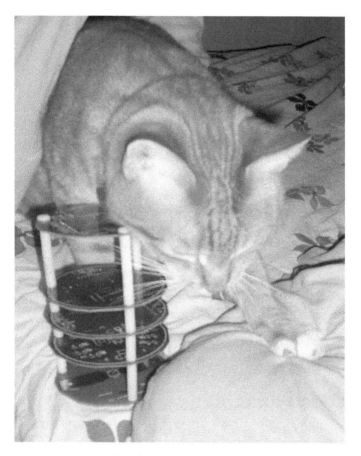

Figure 7-2. *An early build is investigated by a nonspacegoing creature*

Day Job

As this project is a part-time effort, I did not work on Calliope every day. Being the stay-at-home parent, doing freelance work, my volunteer work, the occasional project pitch, networking for my paying work, all those sundry elements take up much of my time. Calliope things took longer than they might have in a commercial or academic environment. This is part of the story, though. We're not just in an era when someone can build a satellite in their basement, we're in a era when he or she can build it as a hobby, not a career.

Goal

All of this is aiming so we—a global "we'" an *anyone* "we"—get to hear the rhythm of the ionosphere and get a sense of just how active space is. What is your mission goal?

If your goal doesn't include "have huge amounts of fun building a satellite," you need to find a better one.

About the Author

Alex "Sandy" Antunes is an astrophysicist who turned to science writing upon realizing the desire to understand the universe doesn't mean you have to be the one to discover everything personally. There's a lot of excellent science out there, and Sandy enjoys bringing it to the world's attention. Sandy recently achieved a professorship at Capitol College's Astronautical Engineering department, which he credits to his NASA work, his solo build of the Project Calliope picosatellite, and his writing for Science 2.0 and via O'Reilly Media.

Have it your way.